T0254926

Regression Analysis in R

Regression Analysis in R: A Comprehensive View for the Social Sciences covers the basic applications of multiple linear regression all the way through to more complex regression applications and extensions. Written for graduate level students of social science disciplines this book walks readers through bivariate correlation giving them a solid framework from which to expand into more complicated regression models. Concepts are demonstrated using real data examples and R software without assuming prior familiarity with R.

- Comprehensive treatment of most common multiple regression applications for researchers in the social sciences.

- Application-based presentation of R code.

- Brief primer on R for the unfamiliar user complete with tips for troubleshooting.

- End of chapter exercises to check your understanding.

Jocelyn H. Bolin is a professor in the Department of Educational Psychology at Ball State University, where she teaches courses on introductory and intermediate statistics, multiple regression analysis, and multilevel modeling to graduate students in social science disciplines. She earned a PhD in educational psychology from Indiana University Bloomington. Her research interests include statistical methods for classification and clustering and the use of multilevel modeling in the social sciences.

Chapman & Hall/CRC
Statistics in the Social and Behavioral Sciences Series

Series Editors
Jeff Gill, Steven Heeringa, Wim J. van der Linden, Tom Snijders

Recently Published Titles

Big Data and Social Science: Data Science Methods and Tools for Research and Practice, Second Edition
Ian Foster, Rayid Ghani, Ron S. Jarmin, Frauke Kreuter and Julia Lane

Understanding Elections through Statistics: Polling, Prediction, and Testing
Ole J. Forsberg

Analyzing Spatial Models of Choice and Judgment, Second Edition
David A. Armstrong II, Ryan Bakker, Royce Carroll, Christopher Hare, Keith T. Poole and Howard Rosenthal

Introduction to R for Social Scientists: A Tidy Programming Approach
Ryan Kennedy and Philip Waggoner

Linear Regression Models: Applications in R
John P. Hoffman

Mixed-Mode Surveys: Design and Analysis
Jan van den Brakel, Bart Buelens, Madelon Cremers, Annemieke Luiten, Vivian Meertens, Barry Schouten and Rachel Vis-Visschers

Applied Regularization Methods for the Social Sciences
Holmes Finch

An Introduction to the Rasch Model with Examples in R
Rudolf Debelak, Carolin Stobl and Matthew D. Zeigenfuse

Regression Analysis in R: A Comprehensive View for the Social Sciences
Jocelyn H. Bolin

Analysis of Intra-Individual Variation: Systems Approaches to Human Process Analysis
Kathleen M. Gates, Sy-Min Chow, and Peter C. M. Molenaar

Applied Regression Modeling: Bayesian and Frequentist Analysis of Categorical and Limited Response Variables with R and Stan
Jun Xu

For more information about this series, please visit: https://www.routledge.com/Chapman--HallCRC-Statistics-in-the-Social-and-Behavioral-Sciences/book-series/CHSTSOBESCI

Regression Analysis in R

A Comprehensive View for the Social Sciences

Jocelyn H. Bolin

CRC Press
Taylor & Francis Group
Boca Raton London New York

CRC Press is an imprint of the
Taylor & Francis Group, an **informa** business

A CHAPMAN & HALL BOOK

First edition published 2023
by CRC Press
6000 Broken Sound Parkway NW, Suite 300, Boca Raton, FL 33487-2742

and by CRC Press
4 Park Square, Milton Park, Abingdon, Oxon, OX14 4RN

CRC Press is an imprint of Taylor & Francis Group, LLC

© 2023 Taylor & Francis Group, LLC

Reasonable efforts have been made to publish reliable data and information, but the author and publisher cannot assume responsibility for the validity of all materials or the consequences of their use. The authors and publishers have attempted to trace the copyright holders of all material reproduced in this publication and apologize to copyright holders if permission to publish in this form has not been obtained. If any copyright material has not been acknowledged please write and let us know so we may rectify in any future reprint.

Except as permitted under U.S. Copyright Law, no part of this book may be reprinted, reproduced, transmitted, or utilized in any form by any electronic, mechanical, or other means, now known or hereafter invented, including photocopying, microfilming, and recording, or in any information storage or retrieval system, without written permission from the publishers.

For permission to photocopy or use material electronically from this work, access www.copyright.com or contact the Copyright Clearance Center, Inc. (CCC), 222 Rosewood Drive, Danvers, MA 01923, 978-750-8400. For works that are not available on CCC please contact mpkbookspermissions@tandf.co.uk

Trademark notice: Product or corporate names may be trademarks or registered trademarks and are used only for identification and explanation without intent to infringe.

Library of Congress Cataloging-in-Publication Data

Names: Bolin, Jocelyn H., author.
Title: Regression analysis in R : a comprehensive view for the social sciences / Jocelyn H. Bolin.
Description: First edition. | Boca Raton : CRC Press, [2022] | Series: Chapman & Hall CRC statistics in social and behavioral sciences | Includes bibliographical references and index.
Identifiers: LCCN 2022003683 (print) | LCCN 2022003684 (ebook) | ISBN 9780367272586 (pbk) | ISBN 9781032257754 (hbk) | ISBN 9780429295843 (ebk)
Subjects: LCSH: Regression analysis. | Social sciences--Statistics.
Classification: LCC HA31.3 .B54 2022 (print) | LCC HA31.3 (ebook) | DDC 519.5/36--dc23/eng/20220414
LC record available at https://lccn.loc.gov/2022003683
LC ebook record available at https://lccn.loc.gov/2022003684

ISBN: 978-1-032-25775-4 (hbk)
ISBN: 978-0-367-27258-6 (pbk)
ISBN: 978-0-429-29584-3 (ebk)

DOI: 10.1201/9780429295843

Typeset in Minion
by Deanta Global Publishing Services, Chennai, India

Contents

Acknowledgments, xi

CHAPTER 1 ▪ Introduction   1

CONTEXTUALIZING CORRELATION AND REGRESSION
ANALYSIS   3

REGRESSION AS PREDICTION   3

REGRESSION AS EXPLANATION   3

CORRELATION, REGRESSION, AND CAUSATION   4

OVERVIEW OF THIS BOOK   5

REFERENCE   6

CHAPTER 2 ▪ Correlation   7

VISUALIZING RELATIONSHIPS   7

UNDERSTANDING COVARIATION   9

SIMPLE LINEAR RELATIONSHIPS: THE PEARSON
PRODUCT MOMENT CORRELATION COEFFICIENT   10

SIGNIFICANCE TESTING FOR THE PEARSON R   11

ASSUMPTIONS OF THE PEARSON R   14

ALTERNATIVE CORRELATIONS: KENDALL TAU AND
SPEARMAN RHO   16

   The Spearman Rho   17

   The Kendall Tau   18

CORRELATION USING R   20

   Correlation Using {stats} Package   20

Correlation Using {Hmisc} Package   21

Matrix Scatterplots Using {Performance Analytics} Package   22

CHAPTER SUMMARY   23

REFERENCES   24

CHAPTER 2: END OF CHAPTER EXERCISES   25

CHAPTER 3 ■ Simple and Multiple Regression   27

SIMPLE LINEAR REGRESSION   27

Ordinary Least Squares (OLS) Regression   28

The Linear Regression Equation   28

Regression Model Fit   30

Multiple R   31

R^2 and Adjusted R^2 31

Standard Error of the Estimate   32

Multiple Regression Analysis   33

OLS Regression Using lm()   35

SUMMARY   38

CHAPTER 3: END OF CHAPTER EXERCISES   39

CHAPTER 4 ■ Assumptions of Multiple Regression   41

STATISTICAL ASSUMPTIONS OF MULTIPLE REGRESSION   41

THEORETICAL ASSUMPTIONS OR 'INTERPRETATIONAL CONSIDERATIONS'   42

The Regression Model Is Theoretically Sound   43

Restriction of Range   43

Absence of Multicollinearity   43

CHECKING ASSUMPTIONS OF MULTIPLE REGRESSION USING R SOFTWARE   44

CHAPTER 4: END OF CHAPTER EXERCISES   54

CHAPTER 5 ■ Dummy Variables and Interactions   55

CATEGORICAL VARIABLES IN REGRESSION   55

DUMMY VARIABLES   57

  A Note on the '0 0' Category   58

USING/INTERPRETING DUMMY VARIABLES IN
A REGRESSION MODEL   58

INTERACTION EFFECTS IN REGRESSION MODELS   61

A NOTE ON INCLUDING MAIN EFFECTS AND CENTERING
FOR PRODUCTS   65

  Centering Predictors Using R   65

CHAPTER SUMMARY   67

CHAPTER 5: END OF CHAPTER EXERCISES   68

CHAPTER 6 ■ Regression vs. ANOVA?   69

ANALYSIS OF VARIANCE   69

ANOVA AS REGRESSION   71

ANOVA OR REGRESSION?   75

CHAPTER 7 ■ Model Comparisons and Hierarchical
Regression   79

WHY COMPARE MODELS?   79

WHAT DOES IT MEAN FOR MODELS TO BE NESTED?   81

MODEL COMPARISONS FOR NESTED AND NON-NESTED
MODELS   81

  Comparisons of Non-Nested Models   82

  R Example of Non-Nested Model Comparison   82

COMPARISONS OF NESTED MODELS   84

  Types of Nested Model Comparison   85

CHAPTER SUMMARY   90

CHAPTER 7: END OF CHAPTER EXERCISES   92

CHAPTER 8 ■ Moderation/Mediation and Regression
Discontinuity   93

EXTENSION 1: MODERATION   93

EXTENSION 2: REGRESSION DISCONTINUITY   96

  Motivating Example   96

Interpreting Treatment Effects in Regression Discontinuity
Design   97

   A Note on the Terminology   97

EXTENSION 3: MEDIATION   102

Baron and Kenny (1986) Requirements for Testing Mediation   103

Tests of Significance for the Indirect Effect   106

END OF CHAPTER SUMMARY   108

RECOMMENDED RESOURCES   108

CHAPTER 8: END OF CHAPTER EXERCISES   109

CHAPTER 9 ■ Non-Linearity and Cross-Validation   111

EXTENSION 4: NON-LINEARITY   111

Variable Transformations for Non-Linearity   112

Transformation Selection   112

What to Do with Negative Values?   113

Pros and Cons to the Transformation Approach   113

Use of Non-Linear Terms   116

Watch out for Multicollinearity!   117

Pros and Cons to the Use of Non-Linear Terms   119

EXTENSION 5: CROSS-VALIDATION   120

Cross-Validation Samples   121

Cross-Validation Procedures   121

END OF CHAPTER SUMMARY   125

CHAPTER 9: END OF CHAPTER EXERCISES   126

CHAPTER 10 ■ Nested Data   127

FIXED EFFECTS MODELING   128

HIERARCHICAL LINEAR MODELING   130

Random Effects and the Tau Matrix   133

HLM Using R Software   134

CONCLUDING COMMENTS ON HIERARCHICAL LINEAR
MODELING   138

SUMMARY   138

RECOMMENDED RESOURCES   139

CHAPTER 10: END OF CHAPTER EXERCISES   140

APPENDIX A: INTRODUCTION TO R, 141

APPENDIX B: NON-PARAMETRIC ANALYSIS BASED ON RANKS, 151

APPENDIX C: R FUNCTION AND PACKAGE INDEX, 155

APPENDIX D: END OF CHAPTER EXERCISE SCRIPT FILE SOLUTIONS, 159

APPENDIX E: GLOSSARY, 169

INDEX, 177

Acknowledgments

To my amazing graduate assistants:
The students really have surpassed the teacher.
Without you I could never have finished this book!

To my two beautiful sons:
You are my world!
But without you this book would have been finished much sooner ☺

CHAPTER 1

Introduction

I N THE MOST GENERAL of terms, the purpose of most academic research is to better understand the world. Sometimes the goal is to determine if certain factors can be manipulated in order to produce desirable outcomes. This is generally the goal of experimental research. The classic example that comes to mind is the randomized controlled drug trial to determine whether new medications should be mass produced. Often, however, the goal is to understand the relationships between factors already existing in the world. In these cases, most often, manipulation and control over factors are not possible. Instead, the goal is to explain relationships between quantities and characteristics as they naturally occur to better understand the world and then potentially use these relationships to predict future performance.

Consider the following scenarios.

Study 1: Correlates of Academic Test Anxiety (TestAnxiety)

Academic test anxiety has been found to significantly impact academic performance. A study of 363 undergraduate students examined the relationship between academic test anxiety, perfectionism, and academic performance. Variables under study included demographic characteristics (age, gender, minority status), undergraduate GPA, mathematics and verbal GRE, physical responses to stress, perceived test threat, study skills, and a four-factor measure of perfectionism. With a better understanding of the relations between these constructs it was hoped that solutions could be presented to help students with academic anxiety.

DOI: 10.1201/9780429295843-1

Study 2: Mask Attitudes (Mask)

During the COVID-19 pandemic, mask wearing became a very central issue. The CDC advised mask wearing in order to help end the health crisis and protect especially vulnerable populations. Yet, many Americans resisted this advice. A study of 156 undergraduate students aimed to better understand people's attitudes toward mask wearing and how personality characteristics and health diagnoses may impact these attitudes. With understanding of the predictors of mask attitudes and mask anxiety potentially better recommendations can be made to help increase mask compliance for future situations.

Study 3: Understanding Academic Dishonesty: (Cheating)

This study aimed to disentangle serious planned cheating offenses from cheating due to momentary panic. Data were collected on a sample of 155 undergraduate students majoring in Business from a large public university. Particular attention was given to the frequency and severity of and justification for cheating. The researchers also wondered if the severity of cheating offense might be different depending on whether the cheating was planned or due to panic. The goal of this project was to better understand the motives for cheating but to also better categorize cheating offenses in order to inform remediation and consequences.

So, what do these studies have in common? Although these three studies are from very different research areas, their goals can all be similarly aligned. All three studies aim to best explain an outcome of interest through a particular lens or theoretical framework. In a best-case scenario, each of these studies would be interested in the ability to predict the outcome from the characteristics measured. Wouldn't it be great if we could reliably predict academic achievement from demographic characteristics and knowledge of the level of test anxiety a student experiences? And wouldn't it be great if we could reliably predict an individual's attitude toward mask wearing from their age, personality characteristics, and whether they have hearing or vision loss? In a perfect world, these are the goals of using statistical methods to assess relationships.

CONTEXTUALIZING CORRELATION AND REGRESSION ANALYSIS

Correlation and regression methodology allows the researcher to assess relationships between variables. This book will begin by looking at correlation analysis. Correlation analysis is the simplest way of assessing the relationship between two continuous variables. For example, we could use a simple correlation to assess the degree and type of relationship between cognitive test anxiety and academic performance. Correlation analysis will provide the jumping off point for our introduction to regression analysis. Regression analysis is a natural extension of correlation which uses the relationships between variables to allow one variable to be predicted by another variable (or variables). In such a way, regression analysis is not just an analysis of relationship, but also of prediction and explanation.

REGRESSION AS PREDICTION

As will be detailed more in Chapter 2, regression is a predictive analysis. Using a regression analysis, a researcher will create an equation that can be used to predict the desired outcome from a set of measured variables. For example, a regression analysis could be used to create a model predicting math SAT from a student's level of test anxiety, the student's gender, and the student's level of study skills. This model could easily be used in the future to predict math achievement for any student given their gender, level of study skills, and level of test anxiety are known. Although regression analysis can easily provide the information to make such predictions, in the social sciences it is somewhat rare for regression to truly be used for a predictive purpose. In order for a predictive model to be of use in making actual predictions, the predictions it provides need to be relatively accurate. Unfortunately, in the social sciences, due to the difficulty of measurement and complexity of constructs, it is often difficult to create a regression model of high enough predictive accuracy to be useful. This is not to say that it cannot be done, but rather that it cannot be done often.

REGRESSION AS EXPLANATION

In the social sciences, it is far more common to use regression for a second purpose; that of construct explanation. As mentioned in the previous section, it is not often that regression models are of high enough accuracy for use as an actual predictive model. That is not to say, however, that models of lower predictive accuracy cannot be informative. Even models with reasonably low predictive accuracy can provide useful information regarding the theoretical structure of construct relations. It is extremely important, however, knowing the limitations of data collection and the complexities of social constructs, to remember that an explanatory regression model is only as strong as the theory it is based on. It is entirely possible to create a regression model that has a high degree of predictive accuracy and appears to have strong explanatory power but may not be meaningful in the context of theory and real-world explanation.

CORRELATION, REGRESSION, AND CAUSATION

Generally speaking, correlation and regression analysis are not often capable of providing causal statements. The common cry is often heard when first learning correlation and regression analysis, 'Correlation does not mean causation!'. This is generally very good advice. Correlation and regression are generally thought of as *quasi experimental* designs, or designs where random assignment of levels of the independent variable is not possible.* Random assignment is a necessary condition for causal statements to be made. Random assignment helps ensure that independent variable groups begin on an even playing field and are not conflated with nuisance or confounding variables. Some variables, however, are naturally unable to be randomly assigned. Take self-esteem as an example. A researcher cannot simply collect a sample of participants and then tell each participant what their level of self-esteem will be. Rather, the researcher must work with the existing level of self-esteem of each participant. This opens the study design up to internal validity threats that make interpretation more difficult. When a third variable (or more) is uncontrolled, it is impossible to disentangle whether the relationship observed is due to an actual true relationship between the two variables of interest or whether it is due to the third variable.

Also of concern is the directionality of the relationship between variables. Demonstrating that two variables are related does not imply a causal direction between the two. If, for example, a relationship is found between motivation and school absences, is it lack of motivation that caused the absences? Or repeated absence causing a lack of motivation?

---

* It should be noted that sometimes it is possible to use a randomly assigned independent variable in a regression context. If this is the case then yes, causal statements may be possible. Since this is more the exception than the rule in correlation and regression analyses, it is best to proceed from a more conservative stance on causal statements.

Or if a relationship is found between the age a student took algebra and SAT scores, this does not imply that all students should be put in algebra as young as possible.

The take home point, here, is that correlational research can be extremely informative but the researcher does need to be careful not to extend the results beyond what they can actually do. Correlation and regression indicate when relationships are present and allow the researcher to describe them. If further speculation is to be done regarding the mechanisms behind these relationships, the researcher must turn to theory and potentially more advanced statistical techniques (see the final chapter of this text).

OVERVIEW OF THIS BOOK

This book will begin the conversation about relationships with Chapter 2 on bivariate correlation. This provides a good starting point for the conversation as the study of relationships generally extends out of this basic framework. Chapter 3 will extend the concept of relationships into prediction and explanation by introducing simple linear regression and multiple regression. This will allow us to get our feet wet with regression concepts and interpretation before extending the discussion to interpretational issues. Chapter 4 will continue the discussion with multiple regression assumptions, and diagnostic statistics encountered in multiple regression. At this point, readers will have learned the basic multiple regression framework. The rest of the book is devoted to customization options and related methodologies. Chapter 5 extends into the use of categorical predictors and interaction effects allowing for more complex designs to be investigated. Following this discussion, Chapter 6 will briefly move away from regression methodology for a discussion of when to use regression versus when to use ANOVA. Chapter 7 will return to the regression framework and introduce the notion of comparisons among regression models and using hierarchical regression for systematic comparison. The last three chapters will combine the knowledge from the previous chapters to illustrate several extensions of the multiple regression model to different types of research questions and data types. Mediation, moderation, regression discontinuity designs, cross-validation, and methods for nested data will all be discussed.

Throughout the text, examples and syntax walkthrough will be provided using R software. Examples and syntax will be presented for the introductory or casual user of R. No prior familiarity with R is assumed. A primer on basic R use is included in Appendix A for the interested reader.

REFERENCE

Stone, T. H., Kisamore, J. L., Jawahar, I. M., & Bolin, J. H. (2014). Making our measures match perceptions: Do severity and type matter when assessing academic misconduct offenses? *Journal of the Academy of Ethics.* DOI 10.1007/s10805-014-9216-0.

Correlation

C HAPTER 1 DESCRIBED A study about understanding variables affecting academic test anxiety. As a researcher examining the variables affecting academic test anxiety, the following questions may surface: is there a relationship between test anxiety and academic performance? What kind of relationships exist between physical symptoms of anxiety and academic performance? Does higher GPA lead to a lower fear of exam taking? Each of these questions can easily be addressed through the use of correlations. Correlation is a measure of the relationship between two variables. The present chapter will describe several common methods for assessing relationships between variables and provide the tools necessary to analyze relationships using R software.

VISUALIZING RELATIONSHIPS

When considering the relationship between two variables, the easiest way to begin is to look at the relationship visually. A relationship can easily be represented visually through the use of a scatterplot. A scatterplot places one variable on the x axis and one variable on the y axis and each point represents a specific case. This allows the researcher to see the general trend of what happens to one variable as the other variable increases/decreases.

For example, Figure 2.1a shows an example of a positive relationship. As one variable, X, increases, the other variable, Y, also increases. Examples of positive relationships might include the relationship between math SAT scores and verbal SAT scores (the higher the math SAT, the higher the verbal SAT also tends to be), the correlation between ambition and stress tolerance (the more ambitious the individual, the more likely they are to be

DOI: 10.1201/9780429295843-2

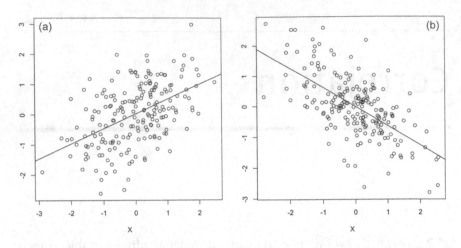

FIGURE 2.1   Scatterplots of a positive and negative relationship between X and Y.

able to tolerate stress), or the correlation between intelligence and school success (the more intelligent the individual, the more likely they are to succeed in school).

Figure 2.1b shows an example of a negative relationship: as one variable, X, increases, the other variable, Y, decreases (or vice versa). Examples of negative relationships might include the relationship between test anxiety and SAT scores (the higher the test anxiety, the lower the SAT scores tend to be), the relationship between test preparation and test anxiety (the more prepared the student is for the test, the less test anxiety they experience), or the relationship between test preparation and physical feelings of anxiety (the less preparation for the test, the more physical feelings of anxiety).

Figure 2.2a shows an example of no relationship between X and Y. For example, there is no relationship between an individual's GPA and the amount of threat they perceive when taking exams. Students of all GPAs experience all levels of test threat. Figure 2.2b shows an example of a non-linear relationship. For example, the amount of support needed across the lifespan tends to be non-linear (more help needed as an infant and small child, less help needed in adolescence and young adulthood, more help needed in old age). This relationship clearly cannot be described by a straight line.

Visualizing relationships is a great place to start when analyzing the relationships between variables as it provides preliminary information about the direction and, to some degree, the strength of the relationship. Visualizing relationships is also important because it gives the researcher

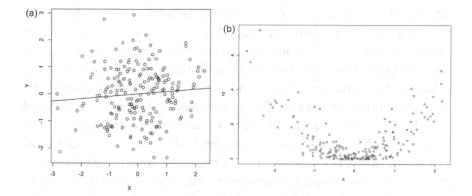

FIGURE 2.2 Scatterplots representing no relationship between X and Y and a non-linear relationship between X and Y.

an idea of the shape of the relationship. Depending on whether linear or non-linear relationships are suspected, different types of analytic procedure may be desired (to be discussed later in this chapter). The discussion will now turn to measuring relationships.

UNDERSTANDING COVARIATION

The discussion of measuring relationships really begins with the *covariance*. *Covariance* is the degree to which two variables vary together.

$$\frac{\Sigma(X-\bar{X})(Y-\bar{Y})}{N-1} \qquad (2.1)$$

X is the value of one variable and Y is the value of the other variable measured in the study. One might notice that this formula is similar to the formula for the variance.[1] Noticing this similarity, we can use what we know about the variance to help us understand the covariance. Recall the formula for the variance:

$$\frac{\Sigma(X-\bar{X})^2}{N-1} \qquad (2.2)$$

In particular, recall that the numerator becomes squared during the computation of the variance (in order to deal with the summation of positive and negative deviation scores). This puts the variance into squared units. Due to the variance being in squared units in comparison with the raw data and the mean being in raw units, the variance is never interpreted. The variance is mathematically useful but due to the squared units, we instead use an interpretable form (the standard deviation), which is in the same raw units as the mean and raw data and allows for easy interpretation.

Back to the covariance. The covariance encounters similar interpretational difficulties, but due to scaling rather than squared units. Both X and Y can be any continuous variable. Take, for example, a correlation between IQ and GPA. IQ and GPA are on completely different scales (IQ: 0–200, GPA: 0–4). Now let's say we want to compare the covariance between IQ and GPA with the covariance between the math SAT and GPA. Math SAT is again on a completely different scale from IQ and GPA (math SAT: 200–800). The resulting two covariances? On completely different scales themselves. There is no standard benchmarking on the covariance because all covariances will be on different scales due to being computed on variables of different scales.

This leaves us with a similar issue to the variance. We have a mathematically informative value that we cannot interpret. We need an interpretable form in order for it to be useful in the applied world. The answer? Standardize the variables! Z scores. When transformed into z scores, variables now have a mean of 0 and a standard deviation of 1. If we turn each variable into a standardized version first (compute z scores for each variable), now both of our variables will be on the same scale. If we then compute the covariance off the standardized values the result is the *correlation*!

SIMPLE LINEAR RELATIONSHIPS: THE PEARSON PRODUCT MOMENT CORRELATION COEFFICIENT

In the social sciences, the most common measure of simple relationship is the *Pearson Product Moment Correlation Coefficient*, better known as the *Pearson r*. The Pearson r is the result of the process described above. If we use z scores in the computation of the covariance, the result is the Pearson r.

$$\frac{\Sigma Z_x Z_y}{N-1} \text{ or } \frac{\text{cov}(XY)}{s_x s_y} \tag{2.3}$$

The standardization of the variables allows the Pearson r to have the same interpretable scale for magnitude and direction of all correlations. The strength or *magnitude* of the Pearson correlation is measured on a scale of –1 to +1. The closer the r value is to +/–1, the stronger the correlation. In other words, the closer the r value is to +/–1, the larger the association or relationship between the variables. An *r* value of 0 is interpreted as no relationship between the variables: the variables are completely unrelated. The sign of the Pearson *r* (positive or negative) provides information on the *direction* of the relationship. A positive value for the Pearson *r* indicates a positive relationship between the two variables: as one variable increases the other variable increases as well (for example, as IQ increases, GPA increases). A negative value for the Pearson *r* indicates a negative relationship between the two variables: as one variable increases, the other variable decreases (for example, as motivation increases, school absences tend to decrease).

In 1988, Cohen published benchmarks for interpreting the magnitude of the correlation coefficient. He suggested $r=.2$ to be a small relationship, $r=.5$ to be a medium effect, and $r \geq .8$ to be a large effect (Cohen 1988). These benchmarks can be incredibly helpful when interpreting the strength and meaningfulness of a significant correlation.

SIGNIFICANCE TESTING FOR THE PEARSON R

Interpreting solely the value of the Pearson *r* allows us to determine the magnitude and direction of the relationship in the sample. In this respect, the Pearson *r* is just a descriptive statistic, like a mean or standard deviation. A significance test for Pearson *r* can be applied, however, to make inferences about the relationship in the population. The Pearson *r* follows a t-distribution with $N - 2$ degrees of freedom. The following formula may be used to test the significance of the Pearson *r*.

$$t = \frac{r\sqrt{N-2}}{1-r^2} \tag{2.4}$$

Using this formula, a *p* value may be obtained for the Pearson *r*. The null hypothesis for this test is $r=0$, or in other words that there is no relationship in the population. Thus, a significant test of the correlation coefficient indicates a non-zero relationship between the two variables in the population.

Examples of Correlation Interpretation Using TestAnxiety Data

Example 1. Is there a relationship between math SAT and verbal SAT?

```
cor.test(TestAnxiety$Math, TestAnxiety$Verbal)
    Pearson's product-moment correlation
```

```
data: TestAnxiety$Math and TestAnxiety$Verbal
```

```
t=11.289, df=374, p-value<2.2e-16
```

```
alternative hypothesis: true correlation is not equal to 0
95 percent confidence interval:

0.4246390 0.5758986
sample estimates:
```

```
      cor
0.5041248
```

According to the Pearson Product Moment Correlation run in R, there is a significant correlation, r=.504, p<2.2e–16 (make sure to review scientific notation as R does not truncate small p values). The relationship between math SAT and verbal SAT is (1) positive, (2) moderate in strength, and (3) significant. Thus, we can say that there is a significant relationship between math SAT and verbal SAT in the population. As verbal SAT increases math SAT also tends to increase.

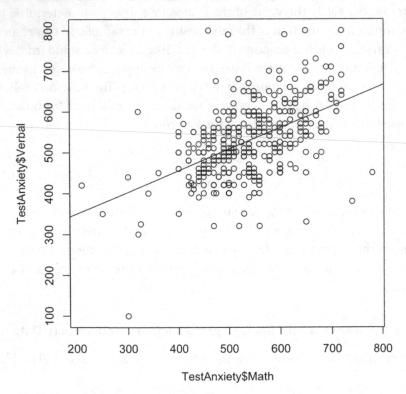

Example 2. Is there a relationship between physical anxiety symptoms and total SAT?

```
>cor.test(TestAnxiety$BStotal, TestAnxiety$TotSAT,
method= "pearson")
        Pearson's product-moment correlation
data: TestAnxiety$BStotal and TestAnxiety$TotSAT
```

```
t=-3.7041, df=377, p-value=0.0002439
```

```
alternative hypothesis: true correlation is not equal to 0
95 percent confidence interval:
-0.2827870 -0.0883231
sample estimates:
```

```
    cor
-0.1873905
```

There is a weak negative significant relationship between physical anxiety symptoms and total SAT ($r = -.187$, $p < .0002$). There is a significant relationship between physical anxiety symptoms and SAT in the population. As total SAT increases, physical anxiety symptoms tend to decrease.

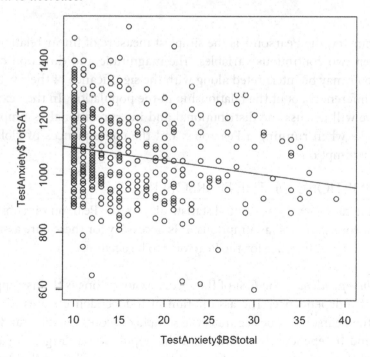

Example 3. Is there a relationship between GPA and perceived test threat?

```
>cor.test(TestAnxiety$GPA, TestAnxiety$PTTtotal,
method="pearson")
        Pearson's product-moment correlation
data: TestAnxiety$GPA and TestAnxiety$PTTtotal
```

```
t=0.089468, df=445, p-value=0.9288
```

```
alternative hypothesis: true correlation is not equal to 0
95 percent confidence interval:

-0.08854214 0.09695144
sample estimates:
```

```
      cor
0.004241137
```

The Pearson correlation between GPA and perceived test threat is not significant (r = .004, p = .9288). There is no relationship in the population. Since the relationship is not significant, we do not discuss the direction of the relationship because the relationship is essentially zero.

In summary, the Pearson *r* is the simplest measure of linear relationship between two continuous variables. The magnitude and direction of the Pearson *r* may be interpreted along with the significance of the *r* value to make inferences about the relationship in the population. In the next section we will discuss the distributional and interpretational assumptions we make when running a Pearson *r* and the consequences of violating these assumptions.

ASSUMPTIONS OF THE PEARSON R

As is the case with all inferential statistical tests, the Pearson *r* is subject to a set of core statistical assumptions. It is necessary for these core assumptions to be met in order for the Pearson *r* to be accurate.

1. *Independence.* The first of these core assumptions is the assumption of independence. The assumption of independence states that, for the Pearson *r* to be accurate, the sample of scores must be randomly and independently drawn from the population at large. Violations of this assumption tend to affect the generalizability of the Pearson

r results: the results from the sample may not accurately generalize to the population of interest. In the social sciences, it is exceptionally difficult to meet this assumption due to ethical and practical constraints with data collection; thus in the social sciences this assumption is likely to be violated a majority of the time. Researchers in this field are cautioned to be aware of this assumption and make all efforts to interpret their results accordingly.

2. *Normality.* The assumption of normality states that for the Pearson *r* to be accurate, the sample of scores needs to come from a population that is normally distributed. Because we do not have information about our population, the only way we can assess this is by looking at the normality of the sample. There are many ways this may be assessed: histograms, skewness and kurtosis statistics, Q–Q plots, statistical tests of normality (e.g. Kolgomorov–Smirnov test). Violations of normality tend to deflate (erroneously reduce) the Pearson *r*. One point of note, however, the Pearson *r* tends to be reasonably robust to violations of normality. Even under moderate cases of non-normality, the Pearson *r* only distorts a small amount (Gayen 1951). Thus, the assumption of normality is not often a big issue for the Pearson *r*.

3. *Scale of measurement.* The assumption of scale of measurement states that for the Pearson *r* to be accurate, the measurement of both variables needs to be continuous (interval or ratio scale). Look back at formula (2.1) for the covariance. What would happen if there were zero variation on the *X* variable? If there were zero variation, $\overline{X}$ would equal *X* for every case making the value of $(X - \overline{X})$ zero! Thus, if there is zero variation on *X*, it does not matter what the distribution of *Y* looks like, the relationship will be zero (see Equation 2.5).

$$\frac{\Sigma(0)(Y-\overline{Y})}{N-1} = 0 \qquad (2.5)$$

Granted, it is unlikely you will encounter a situation where a correlation is desired with a variable having zero variability. Consider, though, what this means about situations of low variability. When zero variability is encountered, the relationship is deflated all the way to zero. When variability is low, the relationship will be deflated as well, just not all the way to zero. In order for relationship to be detected, there must be variation on each variable, otherwise the relationship will be underestimated.

4. *Linearity.* The assumption of linearity states that, for the Pearson *r* to be accurate, the relationship between *X* and *Y* must be linear. In other words, the relationship between *X* and *Y* must be able to be described with a straight line. The reason for this is that the Pearson *r* aims to describe data by describing it with a straight line. If the data cannot be well described by a straight line (for example, if the relationship is curvilinear), the relationship will be underestimated (see Figure 2.2b). The easiest way to determine linearity is through visualization. A *scatterplot* can be used to visualize the relationship between *X* and *Y*. As long as there are no obvious traces of a non-linear relationship, the relationship can be assumed to be linear.

5. *Restriction of range.* Restriction of range refers to the issues encountered when the full range of scores on a variable is not available. For example, what if a college wanted to assess the relationship between college GPA and performance on a college entrance exam? However, the college only accepts students with a college entrance exam above a certain value. Since the full range of scores was not included in this analysis (the low end of the spectrum was not included), the results of the correlation cannot be generalized to the entire range of college entrance exam values.

It should be noted that restriction of range is not considered to be a formal statistical assumption of the Pearson *r* or correlation in general, but rather, should be considered an interpretational issue. Restriction of range will not impact the mathematics of the correlation but rather the way the correlation can be interpreted and generalized.

ALTERNATIVE CORRELATIONS: KENDALL TAU AND SPEARMAN RHO

In the social sciences, the Pearson *r* correlation coefficient is the most commonly used measure of relationship. In light of the assumptions discussed in the previous section, however, there are times when the Pearson *r* is a poor choice.

1. Non-linear relationships. When non-linear relationships are detected, the Pearson *r* is a poor choice of analysis method. As detailed in the previous section, when relationships are non-linear,

the Pearson r will provide a deflated estimate of the relationship due to the test assuming a linear trend.

2. Questionable scale of measurement or variability. When the scale of measurement of variables under study is ordinal or the continuous nature of the variables is under question, the Pearson r may be a poor choice. Similarly, if there is a severe restriction of range or limited variability issue, the Pearson r will have difficulty detecting relationships leading to potentially deflated relationship estimates.

3. Severe normality issues. Although normality is generally of less concern with correlations, instances of several normality violations can be problematic for the Pearson r.

If any of the listed issues are encountered, the researcher may want to use a different analytical method to measure relationships. Non-parametric correlations provide an excellent option in these circumstances. Non-parametric tests are distribution-free statistical tests which do not make assumptions about the parameters or distribution of the population. The Spearman rho and Kendall tau are both rank-based non-parametric measures of relationship which may provide more flexibility when the assumptions of the Pearson r are too strict.

The Spearman Rho

The Spearman rho is a rank based non-parametric measure of bivariate relationship and can be considered an alternative to the Pearson r. The Spearman rho can be achieved by first rank ordering both X and Y from lowest value to highest value (*see Appendix B for further detail*). Then when the ranked data (instead of the raw data) are used to compute the Pearson r, the result is the Spearman rho. The Spearman rho can be interpreted in the same fashion as the Pearson r. Like the Pearson r, the Spearman rho is on a scale from -1 to 1 where the closer to $+1$ or -1, the stronger the relationship between the variables. Spearman rho can also use the same benchmarks for magnitude as the Pearson r (.2 small, .5 medium, .8 strong). The positive or negative on the coefficient is an indication of the direction of the relationship. Significance of the Spearman rho is an indication that the relationship is present in the population.

The Spearman rho can be very useful, especially when data are non-linearly related. In fact, some suggest that if both the Spearman rho and Pearson r are run and the Spearman rho is substantially larger than the

Pearson r, this can be an indication of a violation of the assumption of linearity. *Be careful using this line of thinking, however. There is a fine line between comparing methods for assumption verification and fishing for good results!* It should be noted, however, that while computation of the significance test for the Spearman rho is exact for small samples ($N < 25$) the Spearman rho significance test can only be approximated beyond this value. Thus, Spearman rho is more often suggested for use in small samples.

The Kendall Tau

The Kendall tau is another rank-based non-parametric measure of bivariate relationship. Unlike the Spearman rho, the Kendall tau has no relation to the computation of the Pearson r. Instead, the Kendall tau is based on *concordant and discordant pairs (see Appendix B for further detail)*. First both of your variables are rank ordered from lowest to highest. Then the data are sorted such that one of your variables is in chronological order. Looking at the second variable, concordant pairs are the number of times a large rank is below a specific rank. Discordant pairs are the number of times a lower rank is below a specific rank. What we are essentially doing is quantifying what is happening to the second variable as we increase the value of the first variable. The final formula for the Kendall tau is

$$\tau = \frac{C - D}{C + D} \qquad (2.6)$$

Where C is the number of concordant pairs and D is the number of discordant pairs.

The Kendall tau is also on a scale of –1 to 1 with the closer to +1 or –1 indicating a stronger relationship between the variables and the sign of the coefficient indicating the direction of the relationship. Significance of the Kendall tau still indicates the presence of a relationship in the population. The significance test for the Kendall tau follows a z distribution making it easy to compute for any sample size. The Kendal tau will provide a more conservative estimate of relationship when compared with the Pearson r and Spearman rho. It is important to note, however, that unlike the Pearson r and Spearman rho, there are no published benchmarks for significance of the Kendall tau, making it somewhat more difficult to interpret.

Examples Interpreting Non-Parametric Correlation Using Cheating Data

Example 1: Spearman rho. Is there a relationship between empathy and sensitivity?

```
>cor.test(Cheating$Empathy, Cheating$Sensitive,
method="spearman")
      Spearman's rank correlation rho
data: Cheating$Empathy and Cheating$Sensitive
```

```
S=509264, p-value=0.02549
```

```
alternative hypothesis: true rho is not equal to 0
sample estimates:
```

```
   rho
0.1794265
```

There is a significant Spearman rho correlation between empathy and sensitivity (r = .179, p < .025). There is a significant, positive weak relationship between empathy and sensitivity. As empathy increases, sensitivity also tends to increase. There is a relationship between empathy and sensitivity in the population.

Example 2: Kendall tau. Is there a relationship between the tendency to make performance approach goals and performance avoidance goals?

```
>cor.test(Perfectionism$PAP, Perfectionism$PAV,
method="kendall")
      Kendall's rank correlation tau
data: Perfectionism$PAP and Perfectionism$PAV
```

```
z=15.549, p-value<2.2e-16
```

```
alternative hypothesis: true tau is not equal to 0
sample estimates:
```

```
   tau
0.5873456
```

There is a significant Kendall tau correlation between performance approach goals and performance avoidance goals (r = .587, p < 2.2e–16). There is a significant, positive relationship between performance *approach and performance avoidance. As performance approach goal tendency increases, performance avoidance goal tendency also increases. There is a relationship between performance approach and performance avoidance in the population.*

CORRELATION USING R

There are many packages in R which can facilitate the running of correlations. This book will suggest the use of the {stats} package, the {Hmisc} package, and the {PerformanceAnalytics} package. See Appendix A for further information on the downloading and installation of packages.

Correlation Using {stats} Package

To run correlation using the stats package, the function call is cor() for just the correlation coefficient or cor.test() for a printout of the full inferential information. Using this function, the user may read a dataset into R, and then call the specific variables of interest in the correlation. For example, using the Cheating.sav dataset we can look at the relationship between empathy and sensitivity. Once Cheating is read into R, the stats package may be used to run a scatterplot, a simple correlation, and the test of inference for the correlation. plot(variable, variable) will create a simple scatterplot. The abline() command places the regression line on the plot if desired.

```
>plot(Cheating$Empathy, Cheating$Sensitive)
>abline(lm(Cheating$Empathy ~ Cheating$Sensitive))
```

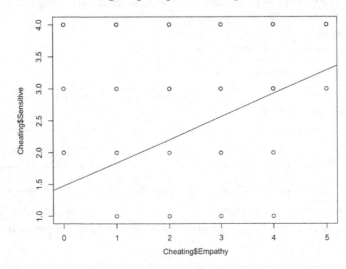

```
> cor(Cheating$Empathy, Cheating$Sensitive)
[1] 0.192244
> cor.test(Cheating$Empathy, Cheating$Sensitive)
      Pearson's product-moment correlation
data: Cheating$Empathy and Cheating$Sensitive
t = 2.4231, df = 153, p-value = 0.01656
alternative hypothesis: true correlation is not equal to 0
95 percent confidence interval:
0.03567706 0.33960003
sample estimates:
   cor
0.192244
```

The {stats} package can also run non-parametric correlations through a similar syntax. Using the same cor.test() function, if we designate the method to be either Spearman or Kendall (instead of Pearson which is the default) the function will produce Spearman rho or Kendall tau.

```
> cor.test(Cheating$Empathy, Cheating$Sensitive,
method = "spearman")
      Spearman's rank correlation rho
data: Cheating$Empathy and Cheating$Sensitive
S = 509264, p-value = 0.02549
alternative hypothesis: true rho is not equal to 0
sample estimates:
   rho
0.1794265
```

```
> cor.test(Cheating$Empathy, Cheating$Sensitive,
method = "kendall")
      Kendall's rank correlation tau
data:  Cheating$Empathy and Cheating$Sensitive
z = 2.2474, p-value = 0.02461
alternative hypothesis: true tau is not equal to 0
sample estimates:
   tau
0.155802
```

Correlation Using {Hmisc} Package

The researcher can use the {Hmisc} package if they wish to run a full correlation matrix instead of just one simple correlation at a time. To do this, the user must first create a matrix in R of the variables to be included in the correlation matrix. The following code will create a matrix which includes columns 1, 6, 9, 16 and 27 from the Cheating dataset.

```
>cheating_cor<- as.matrix(Cheating[c(1, 6, 9, 26, 27)])
>cheating_cor
      Age Empathy Calmness Sensitive Caring
 [1,]   0       2        1         1      0
 [2,]  32       1        2         1      2
 [3,]   0       3        0         1      3
 [4,]  21       2        2         3      3
 [5,]  22       3        4         3      1
 [6,]  24       3        3         3      3
 [7,]  22       3        2         4      4
 [8,]  20       0        0         4      4
 [9,]  22       4        0         3      3
[10,]  22       4        4         4      3
[11,]  23       4        4         4      4
[12,]  22       1        2         4      4
```

Once the desired matrix is created, the `rcorr()` command can be used to run a correlation matrix between all variables in the matrix. The output will yield correlation coefficients in one matrix and p values in a separate matrix below.

```
>rcorr(cheating_cor,type= "pearson")
          AgeEmpathyCalmnessSensitiveCaring
Age        1.00  -0.20     0.09      0.08  0.21
Empathy   -0.20   1.00     0.39      0.19  0.18
Calmness   0.09   0.39     1.00      0.14  0.05
Sensitive  0.08   0.19     0.14      1.00  0.50
Caring     0.21   0.18     0.05      0.50  1.00
n= 155
P
          Age     EmpathyCalmnessSensitiveCaring
Age               0.0113  0.2505   0.3236    0.0093
Empathy  0.0113           0.0000   0.0166    0.0260
Calmness 0.25050.0000              0.0812    0.5196
Sensitive0.32360.0166    0.0812              0.0000
Caring   0.00930.0260    0.5196   0.0000
```

Matrix Scatterplots Using {Performance Analytics} Package

For the visualization of multiple relationships simultaneously (which will become extremely useful when we move to multiple regression) the user can work with the {PerformanceAnalytics} package to obtain a matrix scatterplot. A matrix scatterplot uses the same format as a correlation matrix but instead of correlations, has the scatterplot of each relationship as its entries. Once the {PerformanceAnalytics} is installed, a matrix scatterplot can be run using the `chart.Correlation()` command. The output from `chart.Correlation()` is very useful as it provides scatterplots for linearity as well as the correlations and significance for each bivariate relationship.

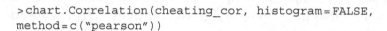

```
>chart.Correlation(cheating_cor, histogram=FALSE,
method=c("pearson"))
```

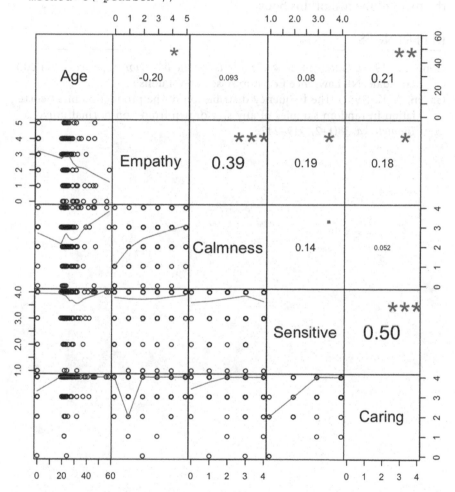

CHAPTER SUMMARY

Correlation analysis is one of the simplest ways to begin to analyze relationships between variables. Depending on the characteristics of your variables and relationship, traditional Pearson correlation or some form of non-parametric correlation may be warranted. Correlations can be used to describe the strength, direction, and significance of the relationship between two variables. R software makes it relatively easy to visualize relationships and run correlations of any type. Chapter 3 will expand upon the relationship framework presented in the current chapter by using these relationships to attempt to predict (or explain) one variable from another. This is the purpose of regression analysis and, subsequently, the focus of the rest of this book!

REFERENCES

Cohen, J. (1988). *Statistical power analysis for the behavioral sciences* (2nd ed.). Hillsdale, NJ: Lawrence Erlbaum Associates, Publishers.

Gayen, A. K. (1951). The frequency distribution of the product moment correlation in random samples of any size drawn from non-normal universes. *Biometrika, 38*(1–2), 219–247.

CHAPTER 2: END OF CHAPTER EXERCISES

Using dataset chapter2ex

1. Run a Pearson correlation between Effort and Performance. (Use any R package.)

   a. Interpret the strength, direction, and significance of this relationship.

   b. Run a scatterplot visualizing this relationship.

   c. Check each assumption for this analysis.

2. Run non-parametric correlations between Distraction and Performance.

   a. Interpret the strength, direction, and significance of this relationship.

   b. Given that the sample size for this relationship is 153, would you prefer to run Spearman rho or Kendall tau for this situation? Why?

Simple and Multiple Regression

CHAPTER 2 BEGAN BY discussing the academic anxiety dataset and sought to describe relationships between the variables of interest (academic test anxiety, physical anxiety symptoms and academic performance). Let us consider why this may be useful. Understanding variables which correlate highly with test anxiety may pave the way for predicting students who may be at risk for academic anxiety and ultimately finding ways to help students who have academic anxiety. However, constructs like academic anxiety are rarely uni-dimensional. In order to truly understand academic anxiety, we will need to use a methodology which draws upon the relationships with all relevant variables and combines this information into a model which can ultimately be used to explain or predict our concept of interest. This brings the discussion to the topic of regression analysis.

SIMPLE LINEAR REGRESSION

Simple linear regression is a natural extension out of correlation. Correlation quantifies the degree of relationship between two variables. Now that relationship can be quantified, simple linear regression essentially uses this relationship to attempt to predict one variable from the other. In particular, as mentioned in the previous chapter, Pearson correlation describes the degree of linear relationship between two variables. Regression builds on this by not only describing the degree of linear relationship between the variables but by creating the equation of the line that

DOI: 10.1201/9780429295843-3

best describes the relationship. This chapter will begin with *simple linear regression*. The purpose of simple linear regression is to explain or predict one variable *Y* (sometimes called the dependent variable, or regressand, or outcome variable) from one variable *X* (sometimes called the predictor, or regressor, or explanatory variable). For the purposes of simplicity, this textbook will generally use the terminology *predictor* for the *X* variable(s) and *outcome* for the *Y* variable.

Ordinary Least Squares (OLS) Regression

Simple linear regression works by finding the *line of best fit* or, in other words, the line that best describes the relationship between *X* and *Y*. In order to do this, we need to define what is meant by 'best fitting'. Several different criteria exist to define a best fitting solution; however, the most commonly used criterion is the OLS or *ordinary least squares* criterion. Ordinary least squares regression defines the line of best fit as the line which minimizes the *sum of squared deviations*. The sum of squared deviations is defined in Equation 3.1 where:

$$\Sigma(Y - Y')^2 \tag{3.1}$$

Y is the outcome variable and *Y'* is the predicted value of *Y* based on the regression model. Decomposing the OLS criterion, we see that it is based on the difference between the predicted values and the actual values. In a perfect world, the difference between the predicted values and the actual values would be zero (perfect prediction!). In reality, prediction is rarely perfect. Therefore, what we are hoping for is the smallest difference between the predicted and actual (also known as a residual) possible. Because some of the difference scores (errors) will be positive and some will be negative, the value will be squared before summing all together. The resulting expression quantifies the total amount of error made by the prediction model. Thus, essentially the goal of the OLS criterion is to place the line of 'best fit' where the sum of squared deviations is minimized, or in other words, where the prediction error is at a minimum (really quite logical if you think about it!).

The Linear Regression Equation

As previously stated, the goal of simple linear regression is to describe the relationship between two variables (one predictor and one outcome) by creating the line which best fits or describes the relationship depicted in the scatterplot. The formula for the line of best fit appears in Equation 3.2 where Y' is the predicted outcome, X is the predictor, a is the *Y-intercept*, and b is the *slope coefficient*.

$$Y' = a + bX \qquad (3.2)$$

This formula might look vaguely familiar to some who have taken algebra in their past. This equation is just the slope intercept form of a straight line ($Y = mX + b$). The definitions of the coefficients of the regression line are also consistent with their algebraic counterparts. The main difference is that there is an accompanying conceptual interpretation for each coefficient above and beyond their mathematical definition.

The Y-intercept. Mathematically speaking, the Y intercept (a in the regression equation) is where the regression line crosses the Y axis. The regression line will cross the Y axis when X (the predictor) is equal to zero. Thus, the Y intercept can be interpreted as the predicted value of Y when the predictor is equal to zero.

The slope coefficient. The slope coefficient (b in the regression equation) quantifies the degree one variable (the outcome variable) changes as the other variable (the predictor) changes. Often this is described as 'rise over run' or 'change in Y over change in X'. In other words, the slope quantifies the amount the outcome will change as we change the predictor variable. Thus, the slope can be interpreted as the amount the outcome changes as we increase the predictor variable by 1. The slope coefficient is generally one of the most important pieces of information from the regression model as it represents the relative importance of the predictor to the outcome.

The regression equation mathematically describes the best fit line relating the predictor (X) to the outcome (Y). The individual pieces of this line, the intercept and slope(s), can be interpreted for information about the relationship between the predictor(s) and outcome. Additionally, the regression equation provides information useful in prediction of the outcome. Once a regression equation is created, if the values of the predictor variable(s) are known the equation can be used to predict the outcome variable in samples where the outcome is unknown. Using the example shown in Figure 3.1, the regression equation predicting math SAT from verbal SAT is $\widehat{Math} = 282.38 + .48(\text{Verbal})$. If we know that an individual's verbal SAT score was 520 we can plug this value into the equation to get a prediction

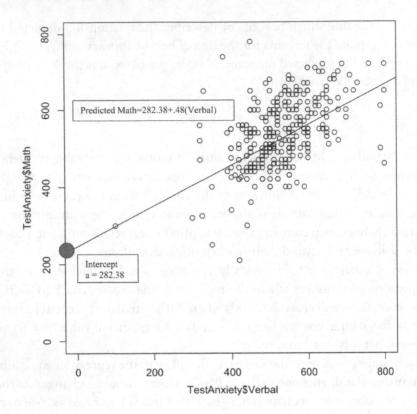

FIGURE 3.1 Scatterplot showing line of best fit, and intercept for test anxiety example.

for the individual's math SAT. $\widehat{Math} = 282.38 + .48(520) = 531.98$. So, with a verbal SAT of 520, we would predict this individual to have a math SAT of 531.98. It is important to remember, however, that predictions are only as good as the model. In order to use a regression equation for predictive purposes, the model should be very strong, accounting for a very high portion of variance explained. The higher the variance explained (closer to 100%) the more accurate the predictions will be.

Regression Model Fit

As previously stated, regression analysis is a statistical approach which creates a linear equation to allow for prediction of one outcome variable from one or more predictor variables. This will allow the researcher to create actual predictions with this model if desired, but also obtain information regarding how effectively the predictor(s) are capable of predicting the outcome. This information, or *model fit*, is often of far more importance to the researcher than the actual ability to use the regression model to predict. Model fit information refers to a set of statistics that quantify the overall strength or explanatory power of the regression model. The name 'model fit' refers to the fact that regression analysis is a basic method of statistical modeling. Statistical models are based in theory and attempt to explain a real-world situation through defining specific relationships between measured variables. Then when we want to see how accurately our statistical model describes the variable or real-world situation of interest, we can look at model fit statistics to see how accurately the model fits or explains the data presented.

Multiple R

Multiple R is a measure of regression model fit defined as the correlation between the predicted outcome and the actual outcome. As such, this value will always be on a scale from 0 to 1 where the closer to 1, the better the fit of the regression model.

R^2 and Adjusted R^2

When you square the multiple R value, you obtain the R^2 model fit statistic. See Equation 3.3. R^2 can be computed as the total variance accounted for by the regression model ($SS_Y - SS_{Residual}$) divided by variance left unaccounted for ($SS_{Residual}$). The value represents the amount of variance in the outcome accounted for by the predictor(s) in the model.

$$R^2 = \frac{SS_Y - SS_{Residual}}{SS_{Residual}} \tag{3.3}$$

R^2 is generally the most commonly used model fit statistic in the social sciences. It is known, however, to be upwardly biased in situations of small sample size or a large number of predictors. Thus, an adjusted version of R squared also exists which adjusts for the sample size and number of predictors used in the analysis. The adjusted R squared can also be thought of as a cross-validation statistic representing the model fit if the model were applied to the entire population instead of a sample.

R^2 and adjusted R^2 can both be interpreted as percent of variance in the outcome variable explained by the predictor variable(s). Thus, R^2 ranges from 0 to 1 with values closer to 1 representing stronger explanatory power. Although some published guidelines do exist for interpretation of R^2 it is generally better to interpret this statistic on a case by case basis considering the number and quality of predictors used as well as explanatory power achieved in previous similar studies.

$$R^2 adj = 1 - \frac{(1-R^2)(n-1)}{n-k-1} \qquad (3.4)$$

Standard Error of the Estimate

If one thinks of multiple R and R^2 as the optimistic measures of model fit, standard error of the estimate can be thought of as the pessimistic measure of model fit. Standard error of the estimate (SEE) is in the metric of the outcome variable and will always represent the average amount of error made when predicting the outcome variable from the predictor(s).

$$SEE = \sqrt{\frac{\Sigma(Y-Y')^2}{N-2}} \qquad (3.5)$$

Regarding which measure of model fit to interpret, largely it does not matter. R^2 tends to be the most common measure of model fit reported due to its easy to interpret scale. SEE can be extremely intuitive when the outcome variable is easily understood or conceptualized (for example, if predicting height in inches or predicting values in dollars); however if the metric is less easily conceptualized (for example scores on a depression or social climate scale) it may be less useful. The main point is that it is impossible for the measures of model fit to disagree with each other since they are all measuring the same thing: just in different ways. Thus it really does not matter which measure of model fit you report, and you do not need to report more than one measure of model fit. They are all essentially the same.

Example 1: Simple Linear Regression Using Test Anxiety Data

How Well Can We Predict Math SAT from Verbal SAT?

```
math.model <- lm(Math ~ Verbal, data = TestAnxiety)
summary(math.model)

Call:
lm(formula = Math ~ Verbal, data = TestAnxiety)
```

```
Residuals:
Min    1Q  Median    3Q   Max
-275.474 -48.665  -4.323  50.184 280.019
Coefficients:
Estimate Std. Error t value Pr(>|t|)
(Intercept) 282.38016  22.90662   12.33  <2e-16 ***
Verbal    0.48356 0.04284   11.29  <2e-16 ***
---
Signif. codes: 0 '***' 0.001 '**' 0.01 '*' 0.05 '.' 0.1 ' ' 1
Residual standard error: 73.73 on 374 degrees of freedom
(110 observations deleted due to missingness)
Multiple R-squared: 0.2541,     Adjusted R-squared: 0.2521
F-statistic: 127.4 on 1 and 374 DF, p-value: < 2.2e-16
```

Begin by looking at the bottom lines of the R output for the overall model significance and the model fit. This regression model predicting math SAT from verbal SAT is significant (F = 127.4, p < .001) and accounts for 25% of the variance in math SAT. Now move up to the coefficients section of the output. The intercept is significant (a = 282.38, p < .001) indicating that the intercept is significantly different from 0. The slope for verbal SAT is also significant (b = .484, p < .001) indicating that verbal SAT is a significant predictor of math SAT: for every one unit increase in verbal SAT, math SAT will increase by .483 points.
You can write the regression equation using the coefficient estimates
Predicted Math = 282.38 + 0.48(Verbal)

Multiple Regression Analysis

Simple linear regression provides a simple way to conceptualize the predictive relationship between a pair of variables. Ultimately, however, simple linear regression can be impractical as a predictive framework due to its simplicity. It is rare that concepts are simple enough to be predicted by one singular aspect. Most concepts (or at least concepts worth studying!) are multidimensional in nature. In other words, most concepts are a composite of multiple dimensions or aspects. Thus, prediction from just one variable seems unreasonable. Luckily, it is very easy to extend simple linear regression to multiple predictors. The resulting analysis is called *multiple regression analysis*. The same statistics and interpretations from simple linear regression still apply to multiple regression making the transition to the multiple predictor framework quite easy.

Example of Multiple Regression Output

Example 2. Predicting Empathy from Sensitivity and Age Using Cheating Data

```
age.model <- lm(Empathy ~ Sensitive + Age, data = Cheating)
summary(age.model)

Call:
lm(formula = Empathy ~ Sensitive + Age, data = Cheating)
Residuals:
Min    1Q Median    3Q    Max
-3.0514 -0.9845 0.0490 1.0155 2.8190
Coefficients:
Estimate Std. Error t value Pr(>|t|)
(Intercept) 2.13984  0.57908   3.695 0.000306 ***
Sensitive 0.39530  0.14668   2.695 0.007830 **
Age     -0.03348  0.01186 -2.823 0.005401 **
---
Signif. codes: 0 '***' 0.001 '**' 0.01 '*' 0.05 '.' 0.1 ' ' 1
Residual standard error: 1.375 on 152 degrees of freedom
Multiple R-squared: 0.08492,    Adjusted
R-squared: 0.07288
F-statistic: 7.053 on 2 and 152 DF, p-value: 0.001177
Standardized Coefficients::
(Intercept)  Sensitive     Age
0.0000000  0.2097793 -0.2197076
```

Like the simple linear regression, we will begin with the bottom of the output and review the overall model significance and model fit. The model using sensitivity and age to predict empathy is significant (F = 7.053, p < .001) and accounts for 8% of the variance in empathy. Moving up to the coefficients section of the output, the intercept is significant (a = 2.14, p < .001) indicating that the intercept is significantly different from zero. The slope for sensitivity (b = .394, p < .01) and the slope for age (−.033, p < .01) are both significant as well indicating that sensitivity and age both significantly predict empathy. The positive slope for sensitivity indicates that empathy increases as sensitivity increases. The negative slope for age indicates that empathy decreases as age increases.

You can write the prediction equation from the coefficient estimates
Predicted Empathy = 2.13 + 0.39(Sensitivity) + −0.033(Age)
If we look at the standardized coefficients we can see that sensitivity and age are both of very similar strength. One would not be considered a stronger predictor than the other.

In summary, OLS regression is a statistical procedure that can be used to predict or explain a continuous outcome variable from one or more predictor variables. This chapter explained how to use measures of model fit to understand the strength of a prediction model and how to interpret intercept and slope coefficients to understand the relative importance of predictor variables and how to write a regression equation. This chapter also extended simple linear regression into multiple regression as, at the core, these are the same model. Chapter 3 will delve deeper into multiple regression to explain the assumptions and interpretational issues associated with prediction using multiple variables. At this time, we will turn the discussion to different R options for running simple and multiple regression.

OLS Regression Using lm()

The most common function in R which can run OLS regression is the lm() function in the {stats} package. The lm() function takes the form lm(formula, data). Here, function refers to the regression model we want to run. Data is the dataset you are using.

For example, using the Cheating dataset, we can run a simple linear regression model predicting Empathy from Sensitivity. The formula is Empathy ~ Sensitive. This roughly translates to Predicted Empathy = $a + b$(Sensitive). Since these variables are in the Cheating dataset, data = Cheating.

When we run this function call in R, we get the following output.

```
sensitive.model <- lm(Empathy ~ Sensitive, data =
Cheating)

Call:
lm(formula = Empathy ~ Sensitive, data = Cheating)
Coefficients:
(Intercept)  Sensitive
1.4782    0.3623
```

This output is very minimal, only presenting the slope and intercept of the regression line. The summary() function can be used to obtain more detailed information about the present model.

```
sensitive.model <- lm(Empathy ~ Sensitive, data =
Cheating)
summary(sensitive.model)

Call:
lm(formula = Empathy ~ Sensitive, data = Cheating)
```

```
Residuals:
Min    1Q Median    3Q    Max
-2.92723 -0.92723 0.07277 1.07277 2.43503
Coefficients:
Estimate Std. Error t value Pr(>|t|)
(Intercept) 1.4782   0.5414  2.730 0.00707 **
Sensitive  0.3623   0.1495  2.423 0.01656 *
---
Signif. codes: 0 '***' 0.001 '**' 0.01 '*' 0.05 '.' 0.1 ' ' 1
Residual standard error: 1.406 on 153 degrees of freedom
Multiple R-squared: 0.03696,     Adjusted
R-squared: 0.03066
F-statistic: 5.872 on 1 and 153 DF, p-value: 0.01656
```

The summary output provides much more detailed output about the regression model including model fit statistics, significance tests, and residuals summary.

Running a multiple regression using the lm() function works in exactly the same fashion as running a simple linear regression. The model itself gets defined in the formula part of the function call. For example, using the Perfectionism data, we could run a multiple regression model predicting self esteem from Positive Emotions (PE_EMO), Engagement (ENG), and Self Reflective Metacognition (SS_META). The formula for this would be

```
esteem<- lm(SE~P_EMO+ENG+SS_META, Perfectionism)
```

If again we want the more detailed output, we should ask for the output summary.

```
summary(esteem)

Call:
lm(formula = SE ~ P_EMO + ENG + SS_META, data =
Perfectionism)
Residuals:
Min    1Q Median    3Q    Max
-3.0872 -0.5088 0.1140 0.5621 1.8832
Coefficients:
Estimate Std. Error t value Pr(>|t|)
(Intercept) 1.78599  0.39626   4.507 8.90e-06 ***
P_EMO    0.52597  0.06567   8.009 1.62e-14 ***
ENG      0.15329  0.05522   2.776 0.00579 **
SS_META   0.02451  0.05746   0.427 0.66994
---
```

```
Signif. codes: 0 '***' 0.001 '**' 0.01 '*' 0.05 '.' 0.1 ' ' 1
Residual standard error: 0.8259 on 360 degrees of freedom
(3 observations deleted due to missingness)
Multiple R-squared: 0.2039,     Adjusted R-squared: 0.1973
F-statistic: 30.73 on 3 and 360 DF, p-value: < 2.2e-16
```

Notice that the lm() output does not provide standardized beta coefficients. Recall that having standardized beta coefficients is of particular use when interpreting multiple regression as it allows for comparison of magnitude of the slope coefficients. We can obtain standardized slope coefficients by using the lm.beta() function from the {lm.beta} package.

```
Library(lm.beta)
esteem.beta <- lm.beta(esteem)

Call:
lm(formula = SE ~ P_EMO + ENG + SS_META, data =
Perfectionism)
Standardized Coefficients::
(Intercept)    P_EMO      ENG    SS_META
  0.0      0.39293489 0.14234222 0.02228129
1.0
```

We can even combine the lm.beta() function with the summary() function to obtain full output with standardized coefficients.

```
summary(esteem.beta)

Call:
lm(formula = SE ~ P_EMO + ENG + SS_META, data =
Perfectionism)
Residuals:
Min   1Q Median   3Q   Max
-3.0872 -0.5088 0.1140 0.5621 1.8832
Coefficients:
Estimate Standardized Std. Error t value Pr(>|t|)
(Intercept) 1.78599   0.00000  0.39626  4.507 8.90e-06 ***
P_EMO    0.52597   0.39293  0.06567  8.009 1.62e-14 ***
ENG     0.15329   0.14234  0.05522  2.776 0.00579 **
SS_META   0.02451   0.02228  0.05746  0.427 0.66994
---
Signif. codes: 0 '***' 0.001 '**' 0.01 '*' 0.05 '.' 0.1 ' ' 1
Residual standard error: 0.8259 on 360 degrees of freedom
(3 observations deleted due to missingness)
Multiple R-squared: 0.2039,     Adjusted R-squared: 0.1973
F-statistic: 30.73 on 3 and 360 DF, p-value: < 2.2e-16
```

SUMMARY

In this chapter we discussed the basic multiple regression model and how to run single and multiple regression models using R software. Understanding this model is framework for all subsequent chapters in this book. It is important to understand, however, all models are based on basic assumptions about the data and when multiple predictors are used in one regression model, complex situations can arise. Chapter 4 will bring Chapter 3 into greater context, discussing the assumptions of multiple regression and providing strategies for assessing assumptions using R software.

CHAPTER 3: END OF CHAPTER EXERCISES

Using dataset chapter3ex

1. Run a simple linear regression predicting Demand from Advertising Budget.

   a. Is the overall model significant?

   b. How well does the model fit?

   c. Interpret the model coefficients (intercept/slope). Are they significant?

   d. Write the prediction model.

2. Run a multiple regression predicting Demand from Advertising Budget, Price, and Availability.

   a. Is the overall model significant?

   b. How well does the model fit?

   c. Interpret the model coefficients. What are the strongest predictors of Demand? (Hint: use the `lm.beta()` function.)

Assumptions of Multiple Regression

IN THE PREVIOUS CHAPTER we explored the prediction framework of simple and multiple linear regression. Multiple regression analysis is a statistical analysis used for the prediction or explanation of a continuous outcome by one or more predictor (or explanatory) variables. Results of multiple regression analysis can be used to determine the strength or fit of the regression model and the relative contributions of each predictor. It is important to remember, however, that all inferential statistical tests are subject to a set of core assumptions about the data in order to ensure accuracy of results. It is also important to realize that the use of multiple predictor (explanatory) variables opens up a set of new interpretational issues to consider on top of making sure all assumptions are met. Thus, the focus of this chapter will be on understanding the statistical and what we will call theoretical assumptions for multiple regression and then considering procedures to check these assumptions using R software.

STATISTICAL ASSUMPTIONS OF MULTIPLE REGRESSION

As multiple regression analysis is an inferential statistical procedure, it is subject to a set of core assumptions about the data and distributions of the variables used. It is crucial that these assumptions be met for a regression model to provide accurate results. The two main statistical assumptions for multiple regression are *scale of measurement* and *multivariate normality*.

DOI: 10.1201/9780429295843-4

The scale of measurement assumption states that all variables (outcome and predictors) must be on an interval or ratio scale of measurement. In other words, variables must be continuous. Scale of measurement is easily determined through knowledge of the variables and measures used or a quick glance at frequency/histogram information.

The assumption of multivariate normality states that all variables and all linear combinations of variables must be normally distributed. This assumption is more difficult to assess directly. Instead, researchers look for indicators that multivariate normality is met. If all four of these indicators are met, multivariate normality is generally assumed to be met. These indicators are as follows: *normality of residuals, homoscedasticity, linearity,* and *independence.* Often, instead of talking about the assumption of multivariate normality, these four indicators are listed as statistical assumptions of regression.

Linearity: each predictor must have a linear relationship with the outcome. In other words, the relationship between predictor and outcome cannot be non-linear. This means that weak or non-significant relationships are not a problem. The only issue for linearity is non-linear (curvilinear) relationships. *This can be visually checked using a simple scatterplot or matrix scatterplot to view all relationships simultaneously.*

Normality of residuals: The assumption of normality of residuals states that the residuals (regression errors) must be normally distributed. *This can be visually checked by running a histogram or normal probability plot of the residuals or by looking at a scatterplot of the residuals.*

Homoscedasticity: The assumption of homoscedasticity states that there must be equal variability in errors across the entire spectrum of predicted values. *This assumption is visually checked using a scatterplot of the residuals.*

Independence: The assumption of independence states that residuals must be uncorrelated. Correlated residuals generally stem from non-random or in particular cluster sampling. *This assumption can be checked visually using a residuals scatterplot as well as using diagnostic statistics such as the Durbin Watson.*

THEORETICAL ASSUMPTIONS OR 'INTERPRETATIONAL CONSIDERATIONS'

The five assumptions just discussed are the core assumptions that are generally considered the statistical assumptions of regression as they will impact the model estimation and actual mathematics of the models. There are also three more important interpretational considerations which should be taken into account when determining the interpretability of a regression model: *theoretical soundness, restriction of range,* and *absence of multicollinearity.* These considerations do not affect the mathematics of the model. For example, they will have no impact on the model variance explained. They will, however, have strong implications for the interpretability of the regression model coefficients, and thus it is very important that these considerations are well understood and considered with each model run.

The Regression Model Is Theoretically Sound

This may seem like it goes without saying, but when the goal of a regression model is conceptual explanation, a regression model is only as strong as the theory it is based on. This means all selected predictors and all selected modeling options should be theoretically justified. This is where the literature review and guiding theoretical models become incredibly important. If a regression model is created where every possible predictor is used and all customizations are included 'just in case', the model may indeed account for a high portion of variance but it may have little theoretical or conceptual interpretation. This is due to potentially nonsensical predictors being used in combination or, as will be discussed in the next section, issues due to predictor overlap which mathematically render coefficients uninterpretable.

Restriction of Range

Restriction of range is the same issue as presented in the chapter on correlation so we will spend little time on it here. As a reminder the issue is that models will have limited generalizability when the range of values on the predictors or outcome does not represent the entire range of values on that variable. When dealing with variables on a restricted range it is important to qualify all interpretations with the range of scores to which those interpretations are applicable.

Absence of Multicollinearity

Checking for absence of multicollinearity is probably the most important interpretational consideration when working with multiple regression models. Multicollinearity refers to potential variance overlap between predictors. Overlap between predictors is problematic for two major reasons.

1. *When predictors have a high degree of overlap it generally means at least one predictor is redundant.* In most academic fields of study, the concepts of interest are multi-faceted. The construct is not likely to be explained by one aspect or variable alone. The goal of multiple regression is to explain these multi-faceted constructs using multiple predictor variables to account for as many facets of the concept as possible. If two predictor variables are highly related to each other, however, they will often explain the same variance in the concept. Having both variables explaining the same piece of variance is unnecessary, thus one of the predictors would be essentially redundant and not actually helping add to the explanation of the outcome.

2. *When predictors have a high degree of overlap only one predictor can claim the common variance leading to nonsensical beta coefficients for the predictor that 'lost'.* As OLS regression is defined, only one predictor can account for each portion of variance in the outcome. When two variables have a high degree of overlap (i.e. one is redundant), still only one of these variables can claim the portion of variance they share. This inevitably means that one variable will take the shared variance while the other variable is left with little variance left to account for. This often leads to strange results like predictors being non-significant when they had strong correlations with the outcome or the sign on the coefficient flipping direction due to the remaining small portion of variance actually being inversely related to the outcome. The ultimate result is that use of predictors with a high degree of overlap can lead to nonsensical results that cannot be interpreted.

To conclude, it is imperative that the assumptions of regression (both statistical and theoretical) be checked every time a multiple regression analysis is run. This will help ensure that the model is sound as far as theoretical framework, model estimation, and model interpretation. In the following sections, procedures for checking statistical assumptions and multicollinearity will be presented along with R code for executing these statistical checks.

CHECKING ASSUMPTIONS OF MULTIPLE REGRESSION USING R SOFTWARE

As detailed in the previous section, the assumptions of multiple regression include scale of measurement, linearity, normality of residuals, homoscedasticity, and independence. It is also very important to check for the absence of multicollinearity. In this section we will go assumption by assumption discussing how to check each assumption using an example from the Perfectionism dataset.

Example 1. Predicting self-esteem from positive emotionality, student engagement, and metacognitive self-regulation. Using Perfectionism data.

```
library(lm.beta)
esteem <-lm(formula = SE ~ P_EMO + ENG + SS_META, data =
Perfectionism)
esteem.beta<-(lm.beta(esteem))
summary(esteem.beta)

Call:
lm(formula = SE ~ P_EMO + ENG + SS_META, data =
Perfectionism)
Residuals:
 Min   1Q Median  3Q   Max
-3.0872 -0.5088 0.1140 0.5621 1.8832
Coefficients:
     Estimate Standardized Std. Error t value Pr(>|t|)
(Intercept) 1.78599   0.00000 0.39626  4.507 8.90e-06 ***
P_EMO   0.52597   0.39293 0.06567  8.009 1.62e-14 ***
ENG     0.15329   0.14234 0.05522  2.776 0.00579 **
SS_META  0.02451   0.02228 0.05746  0.427 0.66994
---
Signif. codes: 0 '***' 0.001 '**' 0.01 '*' 0.05 '.' 0.1 ' ' 1
Residual standard error: 0.8259 on 360 degrees of freedom
(3 observations deleted due to missingness)
Multiple R-squared: 0.2039,Adjusted R-squared: 0.1973
F-statistic: 30.73 on 3 and 360 DF, p-value: < 2.2e-16
```

The model is significant (p <.001) accounting for 20% of the variance in self-esteem. Positive Emotionality and Student Engagement are significant positive predictors of self-esteem with the strongest predictor being Positive Emotionality ($\beta = .39$).

1. Checking the Assumption of Linearity

To check linearity, we can use the same procedure described in Chapter 2 to create scatterplots of the relationships between the predictor(s) and the outcome. First, it is helpful to begin by making a matrix of the variables used in the analysis. We can use this matrix to create a correlation matrix if desired and a matrix scatterplot to help us assess the linearity assumption.

```
library(Hmisc)
library(Performance Analytics)
esteem_matrix <- as.matrix(Perfectionism[c(9, 6,14, 19)])
esteem_scatterplot <- chart.Correlation(esteem_matrix)
esteem_scatterplot
```

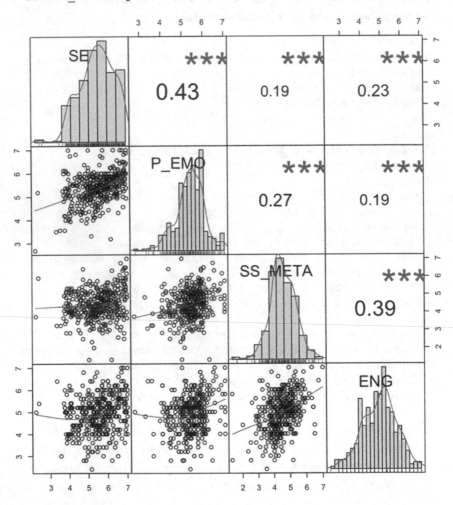

This function call to create a matrix scatterplot also provides information on distribution shape and correlations between variables.

Focusing on the three scatterplots which represent relationships between each predictor and the outcome (self-esteem), we see no evidence of non-linear relationships. Because there is no evidence of non-linearity, we can assume the assumption of linearity is met.

2. *Checking Normality, Homoscedasticity, and Independence*

The rest of our statistical assumptions involve the model prediction errors or *residuals*. The residuals are the difference between the predicted values and the actual values and thus allow us a way to look at assumptions concerning the combined impact of the predictors on the outcome.

We can start by obtaining residuals from the regression model using the call resid(modelname). Once we have obtained the residuals, we can use a number of different methods to check if the model residuals are normally distributed, homoscedastic, and independent.

We could start by viewing a histogram or a Q–Q plot of the residuals to visually check normality.

```
esteem.residuals<-resid(esteem)
hist(esteem.residuals)
```

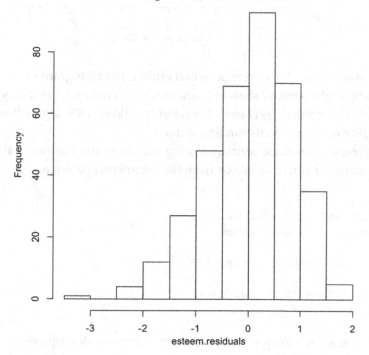

qqnorm(esteem.residuals)

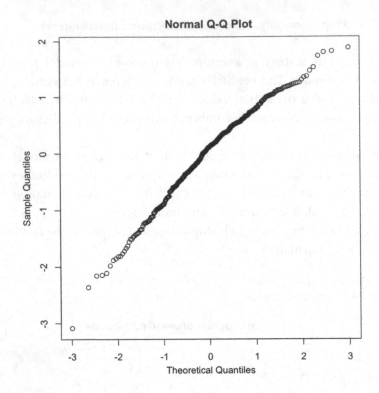

With the exception of one potential outlier, the histogram of residuals has some slight negative skew but otherwise does not look extremely non-normal. The normal Q–Q plot also roughly follows a 45-degree line. The normality of the residuals can be assumed.

We could also check normality using skewness and kurtosis statistics. These statistics can be obtained from the {moments} package.

```
library(moments)
skewness(esteem.residuals)
kurtosis(esteem.residuals)

>skewness(esteem.residuals)
[1] -0.5009555
>kurtosis(esteem.residuals)
[1] 0.1275647
```

Skewness and kurtosis statistics are common descriptive statistics which represent deviation from the normal curve. A value of zero for the skewness and kurtosis statistics represents no deviation from normal (perfectly normal distribution) where non-zero values for skewness and kurtosis indicate some deviation from normality. Generally skewness and kurtosis statistics are considered to be potentially problematic when they reach 1 or –1. Thus in this example, the residuals would not be considered to have a large degree of either skew (–.5) or kurtosis (.12).

We can also use a scatterplot of residuals to check the assumptions of normality of residuals as well as the assumptions of homoscedasticity and independence.

```
plot(esteem.beta)
```

This function call will actually produce four different plots that can be toggled between in R studio. The first of these plots is the scatterplot of the residuals or the *residuals plot*.

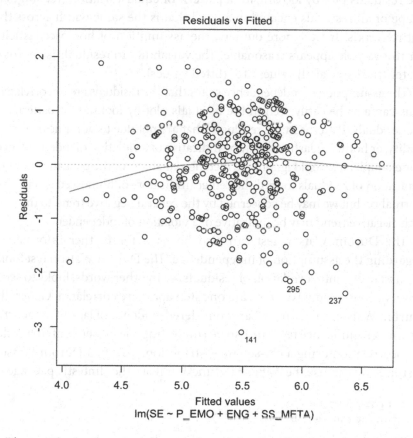

The residuals plot plots the predicted (or fitted) values on the X axis against the standardized residuals on the Y axis. Generation of this plot is one of the most common ways to check assumptions for multiple regression.

Although there are several other ways to check the assumption of normality, it is also easy to see this assumption in the residuals plot. We are looking to see a pile-up of residuals near zero (indicating prediction errors are small) with a larger number of errors as we move further away from zero. Information from the histogram and Q–Q plot of residuals, skewness and kurtosis statistics, and the residuals plot all seem to indicate that the assumption of normality for this model appears reasonable.

As mentioned above, the residuals plot can also be used to assess the assumptions of homoscedasticity and independence. Recall, the assumption of homoscedasticity requires the residuals to be equally varied across the entire continuum of predicted values. This can easily be assessed in the residuals plot by looking for a pattern of equal variation (rectangular shape or all residuals encased in a band that is the same width across the entire X axis. If we ignore outliers, the assumption of homoscedasticity for this sample appears reasonable. The variability in residuals is approximately the same at all values of X (fitted values).

The assumption of independence states that the residuals are uncorrelated. This can also be easily seen in the residuals plot by looking for patterns in the residuals. Patterns in the residuals (that are not due to non-linearity, normality, or homoscedasticity) may be indicative of a violation of independence. For example, we would ignore the rectangular shape of the data and the concentration of residuals toward the center (indicators of homoscedasticity and normality) but we may be concerned by the general negative trend to the plot. This negative trend may be indicative of a violation of independence.

The Durbin Watson test can also be used for further information regarding the assumption of independence. The Durbin Watson test looks for first order autocorrelation of residuals, or in other words, looks to see if observations(residuals) which are one step apart are correlated. Generally, Durbin Watson values near 2 are considered evidence of lack of autocorrelation. A significance test can also be run testing the presence of first order autocorrelation using a chi-square distribution. To run a Durbin Watson test in R we can use the `dwtest()` function from the {lmtest} package.

```
library(lmtest)
>dwtest(esteem.beta)

    Durbin-Watson test
data: esteem.beta
DW=1.881, p-value=0.128
alternative hypothesis: true autocorrelation is greater
than 0
```

The R printout for the Durbin Watson test produced a DW value of 1.881 which is not significant ($p = .128$) indicating a lack of first order auto-correlation of residuals for this model. It is important to note, however, that lack of autocorrelation of residuals does not mean the assumption of independence is met. It just means that this one type of independence violation is not present. It is still of use to examine the model residuals for prominent patterns.

3. Assessing the Presence of Multicollinearity

As mentioned previously, assessing degree of overlap among predictors (multicollinearity), although not a statistical assumption of multiple regression, is of utmost importance. There are many different metrics which can be used to quantify the amount of predictor overlap present in a model. To begin, arguably the simplest way to assess predictor overlap is to use a correlation matrix to assess the strength of correlation between predictors. (Procedure detailed in Chapter 2.)

```
library(Hmisc)
esteem.matrix<-as.matrix(Perfectionism[c(6, 14, 19)])
rcorr(esteem.matrix, type="pearson")
    P_EMO SS_META ENG
P_EMO 1.00  0.27 0.19
SS_META 0.27  1.00 0.39
ENG   0.19  0.39 1.00
n
    P_EMO SS_META ENG
P_EMO   366   366 364
SS_META 366   367 365
ENG     364   365 365
P
    P_EMO SS_META ENG
P_EMO     0e+00  2e-04
SS_META 0e+00      0e+00
ENG   2e-04 0e+00
```

This correlation matrix only displays correlations between predictors (the dependent variable was omitted for simplicity). All correlations between predictors are significant but quite low ranging from $r = .19$ to $.39$ indicating little cause for concern regarding multicollinearity.

It is important to understand, however, that multicollinearity does not just stem from overlap between two predictors. More complex situations of multicollinearity may exist where multiple predictors could combine together to be collinear with another variable. For example, the use of the SAT math, SAT verbal, and SAT combined scores would lead to serious multicollinearity issues because the SAT combined score is made up of the SAT math and SAT verbal. Ideally, issues of this nature would be identified just by examining the theoretical justification of the regression model. However the researcher can also use quantities such as the *variance inflation factor, tolerance, or eigenvalues and condition index* to help identify collinear linear combinations of predictors.

Variance inflation factor (VIF) and tolerance are measures of the amount of variance the predictors combined explain of themselves. VIF is a ratio of the total variance explained by the model to a model using only the one single predictor of interest. A high VIF (10 or above is often used as a benchmark) is an indication that the predictor is highly collinear with the other predictors in the model. Tolerance and VIF are mathematically related such that tolerance = 1/VIF. However tolerance can be interpreted as the proportion of variance in a predictor that is not accounted for the other predictors. Low values of tolerance (.2 or below is often used as a benchmark) are an indication that the predictor of interest is highly collinear with the other predictors in the model.

Eigenvalues and the condition index also provide measures of the degree of overlap among predictors. Model dimensions with very small eigenvalues (this corresponds to very large condition indices) can be indicative of multicollinearity issues. Condition indices are often considered to be problematic at values of 30 or higher. If a large condition index is identified, the variance proportions for each predictor can be checked. Strong proportions (greater than .9) for more than two predictors on the same eigenvalue are indicative that those predictors may have a strong degree of overlap.

VIF, tolerance, and eigenvalues/variance proportions can all be run at once using the ols _ coll _ diag() function in the {olsrr} package.

```
library(olsrr)
ols_coll_diag(esteem)

>ols_coll_diag(esteem)
Tolerance and Variance Inflation Factor
----------------------------------------
Variables Tolerance   VIF
1 P_EMO 0.9187181 1.088473
2    ENG 0.8409850 1.189082
3  SS_META 0.8105596 1.233716
Eigenvalue and Condition Index
```

```
--------------------------------
Eigenvalue Condition Index  intercept      P_EMO
ENG    SS_META
1 3.951081661  1.00000 0.0007583552 0.000946904 0.001563352 1.767998e-03
2 0.021806141 13.46073 0.0810900869 0.186701727 0.031707028 7.092323e-01
3 0.019360104 14.28579 0.0044514528 0.071705855 0.895772542 2.889707e-01
4 0.007752094 22.57608 0.9137001051 0.740645515 0.070957078 2.902573e-05
```

In this example, supporting the results seen in the correlation matrix, there is little reason to suspect multicollinearity issues. Tolerance values are high for all three predictors; VIF is low for all three predictors and, although there are some high condition indices, each dimension is primarily made up of a single predictor.

In summary, it is crucial to check the assumptions of statistical analysis to ensure the fidelity of interpretations. This chapter summarized the assumptions for multiple regression analysis and provided tools to check each assumption using the R software platform. Moving forward, the next few chapters of this text will provide the reader with the ability to tailor models and model comparisons to address specific research questions and data issues. It is important to remember that no matter how complex or well specified a model is designed to be, it is still important to check the assumptions of the model to ensure correct interpretations.

CHAPTER 4: END OF CHAPTER EXERCISES

Using dataset chapter4ex

Begin by running a multiple regression model predicting Demand from Advertising Budget, Price, and Availability. (Same model as exercises from Chapter 3.)

1. Run a matrix scatterplot to check linearity.

2. Check the normality of residuals assumption two different ways.

3. Run a histogram of residuals (you may have already done this for part 2). Use the residuals plot to check the assumptions of homoscedasticity and independence.

4. Run the Durbin Watson statistic and use it to check the assumption of independence.

5. Run tolerance/VIF and variance proportions to check multicollinearity.

Dummy Variables and Interactions

CONGRATULATIONS! AT THIS POINT you have mastered the basic multiple regression framework. Now, the discussion turns from understanding, framing, and interpreting multiple regression to common customization options to allow for more flexibility in modeling and different research questions to be answered. The current chapter will focus on the inclusion of categorical variables and interaction effects in multiple regression analysis. Although these two topics are rather distinct, they make sense to discuss together as they can be viewed as basic building blocks of more complex regression questions such as moderation, regression discontinuity, and model comparison.

We will begin this chapter using motivating examples from the *Mask* dataset. The main research question associated with this dataset will be *'How well can we predict attitudes toward mask wearing based on type of mask worn, personality characteristics, and various demographic factors?'*

CATEGORICAL VARIABLES IN REGRESSION

Up to this point, all variables included in the multiple regression examples in this book have used continuous variables. This makes logical sense as multiple regression makes a scale of measurement assumption which assumes all variables (outcome and predictors) to be measured on a continuum. However, as is often the case, many variables of interest are not truly continuous, but rather are categorical in nature. Common examples of

DOI: 10.1201/9780429295843-5

categorical variables could be political affiliation (Republican, Democrat, Independent), year in school (freshman, sophomore, junior, senior), or any number of yes/no questions (married/not married; smoker/non-smoker; has children/does not have children). In the present framing example, we will focus on two different categorical variables: vision impairment (yes/no) and face mask usage (cloth/paper/other).

So why all the fuss about scale of measurement? Why does it matter if a predictor is categorical rather than measured on a continuum?* The reasoning lies in the scale of measurement assumption. Recall that multiple regression analysis makes the assumption that the outcome variable and all predictor variables are continuous. What this means is that the mathematics behind the regression model assumes values on a predictor represent continuously increasing amounts of the construct being measured. For example, consider the predictor Age. Age is a continuous variable where 21 is larger than 20 and 22 is larger than 21 and so on. Larger values are indicative of more of the construct and values in between 21 and 22 make sense (a person can be 22 and a half). What if we used a categorical variable instead? The variable vision impairment is coded 1 for not impaired and 2 for vision impaired. In this instance it is representative of yes or no, not more or less. 2 is not more vision impaired than 1. 2 represents presence of impairment, and 1 represents no impairment. There is no continuum in between yes and no. (You can't say they are half impaired.) And the issue becomes even more complex when you realize that the coding is arbitrary! We are not required to use 1 for not impaired and 2 for impaired. We could use 5 for not impaired and 10 for impaired. We could even reverse it and use 2 for not impaired and 1 for impaired! So let us summarize the takeaway points:

- Categorical variables do not have an underlying continuum so you cannot assume a higher value indicates a larger amount of the construct.

- Values in between categorical variable value labels are not interpretable.

- Categorical variable value labels are arbitrary so the 'continuum' they appear to have is entirely false.

* We are only talking about categorical predictors when referring to dummy variables. If the researcher wants to model a categorical outcome variable they must use a different type of regression model (generally logistic regression).

For these reasons, categorical variables need to be given a different treatment when entered into a regression model: a treatment that essentially takes the continuum out of the coding. This is the purpose of creating dummy variables.

DUMMY VARIABLES

First to address a common misconception about dummy coding (a misconception, that, as the author I'll admit I had myself until I took a graduate level regression course). *Dummy coding does not simply refer to the process of assigning numeric value labels to categories of a categorical variable. Dummy coding refers to a very specific and systematic method for assigning a value of 0 or 1 to categories of a categorical variable.* Thus, *this process is not* just saying Democrat = 1, Republican = 2, Independent = 3 and calling it a day!

Dummy variables will:

- Always begin with a value of 0 and end with a value of 1.

- Dummy variables always only have two levels (0 and 1). If a categorical variable has more than two levels, more than one dummy variable will be required.

- For any categorical variable with *k* levels, only *k – 1* dummy variables are necessary to fully account for the variable.

Example 1: Vision Impairment

The variable Visionloss from the *Mask* dataset has two levels: 1 = no, not visually impaired, 2 = yes, visually impaired. As discussed above, if the variable were entered into a regression model exactly as is, the model coefficients would be estimated assuming a continuum between 1 and 2 with an interpretation assuming that 2 implies more of the construct (vision impairment) than 1. To remedy the situation we will create a dummy variable for vision impairment that changes the scale from 1 and 2 to 0 and 1. Since this is just a two level variable this is easy: one group gets a value of 0 and the other gets a value of 1. To make the decision of which level to assign a value of 1, conventionally, the group with a value of 1 will also be the name of the new variable. So, if we name the variable vision_impaired, this would imply that 1 = vision_impaired and 0 = not vision impaired.

Example 2: Face Mask Usage

The variable MaskUsed has three levels: 1 = cloth mask, 2 = paper mask, and 3 = other. Like the previous example, this variable cannot simply be entered into the regression model as it is. Instead, it will need to be dummy coded. The process is not quite as simple as for the vision impairment variable, however, as we cannot just say 0 = other, 1 = cloth mask, 2 = paper mask. Dummy variables are always on a scale of 0 and 1. Thus, dummy coding this variable will require the use of multiple dummy variables. As stated above, a variable with k levels will require $k - 1$ dummy variables. Thus, since mask type has three levels, dummy coding will require two variables. This could be done, for example by creating a dummy variable for cloth_mask (1 = yes, 0 = no) and a dummy variable for paper_mask (1 = yes, 0 = no). A third dummy for other is not necessary because if a case is coded 0 for cloth_mask and 0 for paper_mask they must have specified other. In other words, the third dummy variable is completely collinear with the combination of the other two dummy variables.

A Note on the '0 0' Category

As seen in the example above, when a categorical variable has more than two levels, multiple variables are required and a '0 0' category is always assumed. This '0 0' category is generally referred to as a *reference category*. This is a great name for this category as it is the category the other dummied categories will get compared to in the regression model (foreshadowing interpretation). The researcher has complete control over which category is the reference category if they choose to create the dummy variables themselves. This is generally recommended, especially if there are specific logical comparisons to be made among dummied categories.

USING/INTERPRETING DUMMY VARIABLES IN A REGRESSION MODEL

Once coded properly, dummy variables are just like any other predictors used in a regression model. There are just a few additional pieces to the interpretation for a dummy variable which can add to the information gained. Let us work from the following example predicting mask frustration from vision impairment and the Big Five personality characteristics (conscientiousness, extroversion, agreeableness, openness, and neuroticism).

```
Frustration_Model <- lm(Frustration~Vision_Impaired+BFI_
Conscient+BFI_Extro+BFI_Agree+
   BFI_Open+BFI_Neurot, Mask)
summary(lm.beta(Frustration_Model))
```

```
Call:
lm(formula = Frustration ~ Vision_Impaired + BFI_Conscient +
BFI_Extro + BFI_Agree + BFI_Open + BFI_Neurot, data = Mask)
Residuals:
Min    1Q  Median    3Q    Max
-2.53178 -0.50676 0.04314 0.55369 2.24892
Coefficients:
  Estimate Standardized Std. Error t value Pr(>|t|)
(Intercept)   4.450159   0.000000  0.560493  7.940 5.09e-
13 ***
Vision_Impaired 0.104336   0.051090  0.157495  0.662
0.508721
BFI_Conscient  -0.041717  -0.139177  0.024990 -1.669
0.097208.
BFI_Extro   0.033335   0.101004  0.027623  1.207 0.229480
BFI_Agree   -0.088341  -0.315458  0.022134 -3.991 0.000104
***
BFI_Open    0.003855   0.015767  0.019116  0.202 0.840480
BFI_Neurot  0.072709   0.309979  0.018065  4.025 9.15e-05 ***
---
Signif. codes: 0 '***' 0.001 '**' 0.01 '*' 0.05 '.' 0.1 ' ' 1
Residual standard error: 0.9053 on 145 degrees of freedom
(4 observations deleted due to missingness)
Multiple R-squared: 0.2057,      Adjusted R-squared: 0.1728
F-statistic: 6.257 on 6 and 145 DF, p-value: 7.236e-06
```

First, reviewing the model fit statistics at the bottom, we can see that this regression model is significant ($F = 6.257$, $p < .001$) and accounts for 20% of the variance in mask frustration. The only significant predictors of mask frustration were Agreeableness ($\beta = -.315$, $p < .001$) and Neuroticism ($\beta = .309$, $p < .001$). Thus, being less agreeable and more neurotic was predictive of mask frustration. The dummy variable, Vision_Impaired, was not significant. So this tells us several things.

1 The variable Vision_Impairment was not a significant predictor of mask frustration.

2. The unstandardized estimate of the dummy variable can tell us the difference between the groups. Thus $b = .104$ indicates that being vision impaired (coded 1) was .104 higher in mask frustration than not being vision impaired.

3. Dummy variables can be interpreted like t-tests. We already know that being vision impaired leads to higher mask frustration. However, Vision_Impairment being non-significant tells us that the difference in mask frustration between vision impaired and non-impaired individuals is not significant.

4. Inclusion of a dummy variable tells us if there are significant intercept differences between the regression models for the categories of interest. Here, Vision_Impairment being non-significant tells us that the intercept for the regression line for vision impaired is not significantly different from the intercept for the regression line for non-vision impaired. This perspective on dummy variables may not seem as relevant as the previous information in points 1–3; however for certain types of data situation and research question this information can be extremely important. We will return to this idea in Chapter 7 when we discuss regression discontinuity designs.

CREATING DUMMY VARIABLES USING R

Dummy coding in R can be done from the base R package (no additional packages necessary). All that is necessary is to create a new variable using an `ifelse()` statement. The if else statement will code the variable as 1 if one condition is satisfied and 0 if a different condition is satisfied.

DUMMY VARIABLE WITH TWO LEVELS

For example, using the Mask dataset we can create the Vision_Impaired variable from the first example as follows:

```
Mask$Vision_ Impaired<- ifelse(Mask$Visionloss==2, 1, 0)
```

This code will make a new variable (Vision_Impaired) which, when its value is 2 (yes visually impaired), will be coded as 1 and otherwise will be coded 0.

DUMMY VARIABLE WITH 3+ LEVELS

We can apply the same code to a create a dummy variable with three or more levels. For example, we can create dummy variables to represent the MaskUsed variable from the second example. Instead of using just one `ifelse()` statement, though, we will use `ifelse()` to create two dummy variables.

```
Mask$Paper<- ifelse(Mask$MaskUsed==1, 1, 0)
Mask$Cloth<- ifelse(Mask$MaskUsed==2, 1, 0)
```

This code will create one dummy variable named Paper (coded 1 for Paper Mask and 0 for not Paper Mask) and a second dummy variable for Cloth (coded 1 for Cloth and 0 for not Cloth). The use of these two dummy variables together assumes the third category (other) as the reference category. If the researcher instead wanted to make Paper the reference category, the following code could be used instead.

```
Mask$Cloth<- ifelse(Mask$MaskUsed==2, 1, 0)
Mask$Other<- ifelse(Mask$MaskUsed==3, 1, 0)
```

Recall that a third variable is not necessary as only $k - 1$ dummy variables are necessary to represent k levels.

INTERACTION EFFECTS IN REGRESSION MODELS

If you have encountered interactions in ANOVA before, the concept is exactly the same in regression. An interaction is when the effect of one variable (on the outcome) differs depending on the level of another variable. And this combined effect is explained variance above and beyond what the main effects alone are capable of explaining. *Thus in regression terms, an interaction occurs when the effect (or slope) of one predictor on the outcome differs depending on the level of another predictor.* There are two main differences between ANOVA and regression as far as interactions are concerned.

1. Interactions in regression are not assumed to be present. The researcher must intentionally enter the interaction into the model.

2. Interactions in ANOVA are always between categorical variables. Interactions in regression can be between either categorical or continuous variables or a combination thereof (hence beginning with a discussion on dummy variables).

As noted above, in multiple regression analysis, interactions must be purposefully modeled into the regression equation. This is easily done as the interaction is simply the product of the interacting variables. As we move further, let us consider the following motivating example. This time from the *Coachingstress* dataset.

In a sample of NCAA college basketball coaches: in addition to demographic characteristics, is there an interaction between school division (type I or III) and years of coaching experience on task-based stress?

In this example we are building a multiple regression model where two of the predictors (years of coaching experience and school division) are assumed to interact in their effect on the outcome variable. This would be represented in the model as the product of coaching experience and school division, and this product would have its own slope in the model. Thus, the resulting regression model would be:

$$\widehat{TStress} = a + b_1(age) + b_2(gender) + b_3(Experience) + b_4(DivI) + b_5(ExpXDivI)$$

In this model we first make sure that the variable Division is dummy coded to DivI where 0 = Division I, 1 = Division III. Using the code as.factor() makes sure R views this variable as a categorical variable.

```
CoachingStress$DivI <- ifelse(CoachingStress$Division==2,
1, 0)
as.factor(CoachingStress$DivI)
```

Then the rest of the interaction creation is easily taken care of by the lm() function in R. The interaction product is simply defined as Experience*DivI.

```
library(lm.beta)
library(sjPlot)
library(sjmisc)
library(ggplot2)
Mod_Model <- lm(Task~Age+Gender+Experience+DivI+Experience
*DivI, CoachingStress)
Mod_Modelbeta<-summary(lm.beta(Mod_Model))
```

```
Call:
lm(formula =Task~Age+Gender+Experience+DivI+Experience *
DivI, data=CoachingStress)
Residuals:
Min    1Q Median    3Q    Max
-15.3313 -5.1977 -0.6798  4.7875 18.1721
Coefficients:
 Estimate Standardized Std. Error t value Pr(>|t|)
(Intercept)   19.81054   0.00000  5.56638  3.559 0.000785 ***
Age        0.22965    0.31562  0.15318  1.499 0.139641
Experience  -0.10317   -0.12415  0.19154 -0.539 0.592337
Gender     -1.87183   -0.09934  2.47105 -0.758 0.452042
DivI        2.81199    0.17553  3.72763  0.754 0.453908
Experience:DivI -0.41180   -0.49108  0.21229 -1.940 0.057635.
---
Signif. codes: 0 '***' 0.001 '**' 0.01 '*' 0.05 '.' 0.1 ' ' 1
Residual standard error: 7.703 on 54 degrees of freedom
(7 observations deleted due to missingness)
```

```
Multiple R-squared: 0.1592,      Adjusted
R-squared: 0.08136
F-statistic: 2.045 on 5 and 54 DF, p-value: 0.0868
```

The first point to notice is that none of the main effects included in the model, Age, Gender, Experience, or DivI significantly relate to task-based stress. The interaction between Experience and DivI has a p value of .057. (Note that some researchers would round this up to $p = .06$ and treat as non-significant. Others would interpret it as approximately $p = .05$. The treatment as significant or not in this context is very much researcher dependent. For the sake of illustration, this effect will be considered significant for this example.)

The relationship between task-based stress and experience is negative overall ($\beta = -.124$) but, as previously noted, is non-significant. This non-significance of the main effect, however, may be explainable by the interaction with DivI. A plot can be created using the {sjPlot} package to help better understand this effect. The plot _ model() command is a very simple way to visualize an interaction. The function just needs to know what type of model you want; here we want predictive or 'pred' and the terms of the model. The terms are the interacting variables, so here Experience and DivI.

```
library(sjPlot)
plot_model(Int_Model, type="pred", terms=c("Experience",
"DivI"))
```

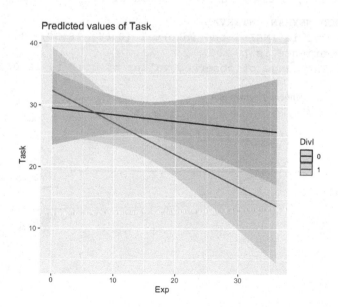

The plot generated depicts the relationship between task-based stress and experience for both division I (coded 1) and division III (coded 0). Depicted in the plot, the slope for division III is weaker than the slope for division I. Thus, the relationship between task-based stress and experience appears to be less strong for division I coaches than for division III coaches.

Creating this plot also helps put dummy variables and interactions into better perspective. We can easily see in this plot that separate regression lines are depicted for division I and division III. With the X axis beginning at 0, it is clear that the intercepts of the two regression lines are very similar (hence the non-significant effect of DivI ($t = .754, p = .45$). However there is a discernable difference in the slopes of the two regression lines coinciding with the significant interaction effect between DivI and Experience ($t = 1.94, p = .05$).

In addition we can obtain Johnson Neyman output from the {interactions} package to further explain this effect. Johnson Neyman intervals help tease apart an interaction by showing what intervals of the observed relationship have significant slopes. Using the johnson_neyman() function, the model needs to be identified, and the two interacting variables need to be defined. There are also other options that can also be specified (alpha level for significance, whether a plot is created).

```
library(interactions)
johnson_neyman(Int_Model, Experience, DivI, vmat=NULL,
alpha=0.05,
plot=TRUE, title="Johnson-Neyman plot")
```

JOHNSON-NEYMAN INTERVAL
When DivI is INSIDE the interval [0.64, 18.37], the slope
of Experience is p<.05.
Note: The range of observed values of DivI is [0.00, 1.00]

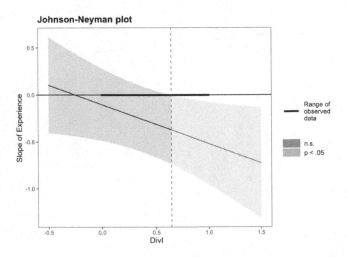

According to the Johnson Neyman interval, the relationship between experience and task-based stress (the slope) will be significant when division is between .64 and 18.37. *Taking into consideration that division was only coded 0 and 1 in this data, this tells us that division changes the relationship between experience and task-based stress such that the relationship is only significant for division I coaches (coded 1).*

A NOTE ON INCLUDING MAIN EFFECTS AND CENTERING FOR PRODUCTS

Before moving forward, it is very important to discuss the way models with interactions should be built. Recall that an interaction is defined as the combined effect of two variables above and beyond each of these variables considered separately. Thus, due to the nature of interactions, an interaction cannot have the desired interpretation if the main effects are not also accounted for in the model. This means that when an interaction is to be modeled into a regression model, the main effects of the interacting variables should also be included in the model. Inclusion of the main effects allows the interaction to truly represent the unique combined effect of the independent variables. However, inclusion of the main effects along with the interaction also has an unintended consequence. The interaction, being a product of the two main effects, will be highly correlated with each of the main effects. Thus, often interactions introduce multicollinearity issues. This is not a reason not to include the main effects. Instead, mean centered variables can be used to create the interaction product which tends to reduce this multicollinearity issue. The reduction in multicollinearity is most prominent when both variables are centered; however if only centering one predictor is desired (for example, if one interacting variable is categorical) even just centering one variable will be effective.

Centering Predictors Using R

A procedure very similar to that of creating dummy variables can be used to center predictors. This can be done just in the base R package. For the coaching stress example, it makes the most sense to just center Experience, since Div1 is categorical. This can be accomplished with the following code.

```
CoachingStress$cExperience<-(CoachingStress
$Experience -    mean(CoachingStress$Experience))
```

Once the predictor(s) are centered, the regression model can be rerun using the centered predictors to create the interaction effect.

```
cInt_Model <- lm(Task~Age+Gender+cExperience+DivI+cExp
erience*DivI, CoachingStress)
summary(cInt_Model)

Call:
lm(formula
= Task ~ Age + Gender + cExperience + DivI + cExperience *
DivI, data = CoachingStress)
Residuals:
Min    1Q Median    3Q   Max
-15.3313 -5.1977 -0.6798  4.7875 18.1721
Coefficients:
    Estimate Std. Error t value Pr(>|t|)
(Intercept)     18.2891   7.0703  2.587  0.0124 *
Age            0.2296   0.1532  1.499  0.1396
Gender        -1.8718   2.4710 -0.758  0.4520
cExperience    -0.1032   0.1915 -0.539  0.5923
DivI          -3.2605   2.0180 -1.616  0.1120
cExperience:DivI -0.4118   0.2123 -1.940  0.0576 .
---
Signif. codes: 0 '***' 0.001 '**' 0.01 '*' 0.05 '.' 0.1 ' ' 1
Residual standard error: 7.703 on 54 degrees of freedom
(7 observations deleted due to missingness)
Multiple R-squared: 0.1592,      Adjusted
R-squared: 0.08136
F-statistic: 2.045 on 5 and 54 DF, p-value: 0.0868
```

Notice that the R^2 values for the centered and uncentered versions of this model are identical. This is exactly as it should be. Centering variables only changes the scale of the predictors. So the model fit will remain unaffected. What will change will be the slope coefficients, p values, and degree of multicollinearity. To explore the changes in multicollinearity, we can quickly run the tolerance/VIF values for both models using the procedures described in Chapter 4. Looking at the uncentered model, the tolerance is very low for Age, Experience, DivI, and the interaction, indicating a high degree of multicollinearity. After centering Experience, tolerance/VIF values improve greatly for DivI and the interaction. Although not all multicollinearity issues were resolved (likely there is overlap between Age and Experience) the multicollinearity due to the inclusion of the interaction has improved.

```
library(olsrr)
ols_coll_diag(Int_Model)
ols_coll_diag(cInt_Model)
    >ols_coll_diag(Int_Model)

#Model without centering
Tolerance and Variance Inflation Factor
------------------------------------------
Variables Tolerance    VIF
1       Age 0.3513003 2.846568
2   Experience 0.2930873 3.411953
3    Gender 0.9053856 1.104502
4      DivI 0.2875674 3.477446
5 Experience:DivI 0.2429445 4.116166
>ols_coll_diag(cInt_Model).

#Model with centering
Tolerance and Variance Inflation Factor
------------------------------------------
Variables Tolerance    VIF
1       Age 0.3513003 2.846568
2     Gender 0.9053856 1.104502
3   cExperience 0.2930873 3.411953
4      DivI 0.9812261 1.019133
5 cExperience:DivI 0.6048999 1.653166
```

CHAPTER SUMMARY

The main takeaway points for this chapter are that categorical variables and interactions can be intentionally modeled into regression models and the inclusion of these terms can be very valuable. The proper use of dummy variables can incorporate very important categorical variable information into a regression model that can often times eliminate the need for additional regression analyses to be run and compared. Interaction terms add to the explanatory power of a model by allowing the combined effect of predictors to be examined. When significant, these interaction effects often provide very rich information that greatly enhances the conceptual explanation of a model. In the next chapter we will look at methods of model comparison and discuss how to build sets of models to address specific research questions.

CHAPTER 5: END OF CHAPTER EXERCISES

Using dataset chapter5ex

1. Using the variable, gender (2 = boy, 1 = girl), create a dummy version of this variable called 'boy' where 1 = boy and 0 = girl.

2. Using the variable, grade (1 = elementary, 2 = middle school, 3 = high school), dummy code into two variables 'elementary' and 'middle'.

3. Run a regression model predicting attention from screen time, age, boy, and the interaction of screen time and age.

   a. Is your model significant? How much variance does it account for?

   b. Interpret your predictors. Is your interaction significant? Interpret. Create a plot for your interaction.

CHAPTER 6

Regression vs. ANOVA?

A T THIS POINT IN the text, it is time to take a step back and examine the situation. What have we learned about regression? What kinds of questions can regression answer? Let us take a minute to summarize. Regression models:

1. Explain variance in a continuous outcome using one or more explanatory variables.

2. Can incorporate both continuous and categorical explanatory variables.

3. Can incorporate interactions between explanatory variables.

4. Partitions variance into the variance explained by the explanatory variables and the variance left unexplained by the model.

At this point some may be thinking, 'wait a minute? Isn't that what analysis of variance does?' And you would be correct. Analysis of variance (ANOVA) is not the focus of this text; however at this point it becomes relevant to discuss because ANOVA and regression are necessarily linked. In fact, that is not a correct statement. Regression and ANOVA are, in fact, mathematically the same model.

ANALYSIS OF VARIANCE

Analysis of variance (ANOVA) is an inferential statistical procedure which partitions outcome variance into the variance explained by the model and

DOI: 10.1201/9780429295843-6

the variance left unexplained. Mathematically this partitioning is accomplished through the computation of sums of squares.

Sum of squares between ($SS_B = \Sigma(\bar{X}_j - \bar{X}..)^2$): sum of squares between is a measure of the variance between the groups means around the grand mean. It represents the explained variance, or the variance accounted for by the independent variables (explanatory variables) in the model. In ANOVA, this is a measure of group differences.

Sum of squares error ($SS_E = \Sigma(X_i - \bar{X}_j)^2$): sum of squares error is a measure of the variance within each independent variable group. It represents the variance in the model not explained by the independent variables. In other words, it represents the unexplained model variance or error variance.

Sums of squares total ($SS_B = \Sigma(X_i - \bar{X}..)^2$): sum of squares total is a measure of the total variance in the model. The sum of squares total can be partitioned into the explained variance (sum of squares between) and the unexplained variance (sum of squares error) $SS_B + SS_E = SS_T$.

From here, the sums of squares are turned into mean squares and an F statistic is computed. See equations below.

$$\left(\frac{\text{sum of squares}}{\text{degrees of freedom}} \right) \tag{6.1}$$

$$F = \frac{\text{Mean Square Between}}{\text{Mean Squre Error}}. \tag{6.2}$$

Relating ANOVA back to regression, ANOVA can also be expressed in terms of its *general linear model* equation. The general linear model is the name for the large family of models that describes an outcome with a linear regression equation. This family includes analyses such as correlation, regression, ANOVA, and ANCOVA. The general linear model equation for ANOVA can be expressed as:

$$Y_{ij} = \mu + \alpha + \varepsilon_{ij} \tag{6.3}$$

Where:
Y_{ij} is the outcome variable for the ith individual in the jth group.
μ is the variance in the dependent variable free from the effect of the 'treatment' or independent variable influence.
α is the 'treatment effect' or the effect of the independent variable(s).
ε is the error left unexplained.

One more set of substitutions makes it clear that regression and ANOVA are describing independent/dependent variable relationships in the same way. The μ is essentially analogous to the intercept (*a*) in the regression model. α is analogous to the slope coefficients and ε is analogous to the error term (*e*). Thus, the equation for ANOVA becomes:

$$Y = a + b_1 (\text{independent variable}) + e$$

So can an ANOVA be run using regression? Yes!

ANOVA AS REGRESSION

Let us return to the data used in Chapter 5 of this text. Chapter 5 demonstrated procedures for using categorical variables and interaction effects in regression models using the Mask data. This provides us with the perfect demonstration of the equivalence of regression and ANOVA. In the previous chapter we were interested in the variable MaskUsed which was coded 1 = cloth, 2 = paper, 3 = other. One research question the researchers who collected this data were interested in was 'Is there a significant difference in perceived effort while wearing a mask between individuals who typically wear different types of masks?'

This research question is easily answered using ANOVA. ANOVA results show a significant omnibus test ($F = 3.755$, $p < .05$) indicating significant differences in effort between the three mask types. Tukey pairwise comparisons were run to follow up the significant omnibus effect. Significant differences are found between cloth masks and other ($p < .05$). No other significant differences are detected.

```
anova.table<- aov(MaskEffort~MaskUsed, Mask)
summary(anova.table)
Df Sum Sq Mean Sq F value Pr(>F)
MaskUsed   2  8.69  4.343  3.755 0.0257 *
Residuals  143 165.36  1.156
---
Signif. codes: 0 '***' 0.001 '**' 0.01 '*' 0.05 '.' 0.1 ' ' 1
10 observations deleted due to missingness
```

We could use a regression model to answer the same question (and obtain identical results). This ANOVA, represented in GLM equations, would be:

$$\text{MaskEffort}_{ij} = \mu + \text{MaskUsed} + \varepsilon_{ij}$$

Translating to the regression equation we would obtain the following:

$$MaskEffort = a + b_1 (MaskUsed) + \varepsilon_{ij}$$

And since MaskUsed is a categorical variable with three levels, the final resulting regression equation would involve dummy coding MaskUsed into two variables. Here it makes sense to make one dummy variable for cloth and one variable for paper resulting in the final regression equation:

$$MaskEffort = a + b_1 (cloth) + b_2 (paper) + \varepsilon_{ij}$$

```
regression<- lm(MaskEffort~cloth+paper, Mask)
summary(regression)

lm(formula=MaskEffort~cloth+paper, data=Mask)
Residuals:
Min   1Q Median   3Q   Max
-1.8128 -0.7640 -0.0709 0.8953 2.2302
Coefficients:
  Estimate Std. Error t value Pr(>|t|)
(Intercept)  0.8143   0.2777  2.933 0.00391 **
cloth      0.6429   0.2958  2.173 0.03140 *
paper      0.9985   0.3673  2.719 0.00737 **
---
Signif. codes: 0 '***' 0.001 '**' 0.01 '*' 0.05 '.' 0.1 ' ' 1
Residual standard error: 1.075 on 143 degrees of freedom
(10 observations deleted due to missingness)
Multiple R-squared: 0.0499,     Adjusted
R-squared: 0.03661
F-statistic: 3.755 on 2 and 143 DF, p-value: 0.02573
```

Now compare. The most obvious statistic to compare is the F test for the ANOVA and the regression model. The F tests are identical both being $F = 3.755$ with 2 and 143 degrees of freedom $p = .0257$. Clearly these two approaches are identical! How else can we see this equivalence? How about effect size? The regression has an effect size of $R^2 = .0499$. The function we used for the ANOVA does not immediately produce an effect size so we can run some additional code using the {lsr}package.

```
Library(lsr)
etaSquared(anova.table)
eta.sq eta.sq.part
MaskUsed 0.04990187 0.04990187
```

The eta square effect size for the ANOVA is .0499! Why is the eta squared equivalent to the R squared? It comes down to the formulas for these statistics. As demonstrated in the equations below, R^2 and η^2 are equivalent measures of effect size! The sum of squares regression from a regression model is equivalent to the explained variance (the sum of squares between of the ANOVA model) and the sum of squares residual is equivalent to the unexplained variance (the sum of squares error of the ANOVA model).

$$R^2 = \frac{SS_{Regression}}{SS_{Residual}} = \frac{SS_{Between}}{SS_{Error}} = \eta^2$$

There is one more important point of equivalence to notice but it is less apparent initially. The effect of mask type (the independent variable of the ANOVA) is represented as two dummy variables in the regression. Clearly these represent the same effect. So why are the slope coefficients for paper and cloth not the same as the effect for MaskType in the ANOVA? It is because we are comparing apples to oranges. The effect of MaskType is variance explained by all three levels of MaskType simultaneously. The regression breaks this effect down into variance explained by cloth and variance explained by paper. A more fair comparison is to look at the Tukey pairwise comparison follow up test for the ANOVA.

```
TukeyHSD(anova.table)
>TukeyHSD(anova.table)

Tukey multiple comparisons of means
95% family-wise confidence level
Fit: aov(formula=MaskEffort~MaskUsed, data=Mask)
$MaskUsed
  diff     lwr       upr       p adj
2-1 0.3555865 -0.2630771 0.97425016 0.3640207
3-1 -0.6429205 -1.3435260 0.05768507 0.0792753
3-2 -0.9985070 -1.8684060 -0.12860806 0.0200373
```

The Tukey pairwise comparison compares each pair of groups. 2-1 compares the group coded 2 (paper) to the group coded 1 (cloth). Likewise, 3-1 compares the group coded 3 (other) to the group coded 1 (cloth). Interpreting the Tukey pairwise we learn that there is a significant difference between paper masks and other types in terms of mask effort. No other significant differences were detected.

Given the way we dummy coded MaskType in the regression, the comparisons we are interested in are 3-1 (other versus cloth) and 3-2 (other versus paper). Looking at the p adj column we again do not see the same *p* values are reported for the dummy variables cloth and paper in the regression! The reason is because the Tukey pairwise comparison does pairwise comparisons of means but adjusts the p value for type I error inflation. If we want to see the model as identical to the regression model, we must request an unadjusted pairwise comparison.

```
pairwise.t.test(Mask$MaskEffort, Mask$MaskUsed,
p.adj = "none")

Pairwise comparisons using t tests with pooled SD
data: Mask$MaskEffort and Mask$MaskUsed
1    2
2 0.1756 -
3 0.0314 0.0074
```

The unadjusted p values for the 3-1 and 3-2 comparisons are now clearly identical to the p values produced in the regression model!

As another proof of concept, let us also run a two-way ANOVA as a multiple regression model. Let us look at a second research question involving the Mask data: do gender identity and whether or not you believe mask wearing infringes on your rights have an effect on perceived mask effort? This question can be answered with a factorial two-way ANOVA with main effects for gender identity and infringe and an interaction between gender identityXinfringe.

```
anova.table2 <- aov(MaskEffort~GenderIdentity+Infri
nge+GenderIdentity*Infringe, Mask)
summary(anova.table2)

              Df Sum Sq Mean Sq F value  Pr(>F)
GenderIdentity   1   0.15    0.15   0.169   0.682
Infringe         1  43.52   43.52  47.472 1.67e-10 ***
GenderIdentity:Infringe  1   0.20    0.20   0.220   0.640
Residuals      142 130.17    0.92
---
Signif. codes: 0 '***' 0.001 '**' 0.01 '*' 0.05 '.' 0.1 ' ' 1
10 observations deleted due to missingness
```

The same model could be run using regression with dummy coded variables for GenderIdentity and Infringe and an interaction between GenderIdentityXInfringe.

```
regression2 <- lm(MaskEffort~GenderIdentity+Infringe+Ge
nderIdentity*Infringe, Mask)
summary(regression2)
```

```
Call:
lm(formula = MaskEffort ~ GenderIdentity + Infringe +
GenderIdentity *
Infringe, data=Mask)
Residuals:
Min   1Q Median  3Q   Max
-1.3244 -0.5562 -0.1507 0.6895 2.0125
Coefficients:
          Estimate Std. Error t value Pr(>|t|)
(Intercept)         3.1162   1.6931  1.840  0.0678.
GenderIdentity      0.6979   0.9511  0.734  0.4643
Infringe           -1.1507   0.9220 -1.248  0.2140
GenderIdentity:Infringe -0.2403  0.5128 -0.469  0.6401
---
Signif. codes: 0 '***' 0.001 '**' 0.01 '*' 0.05 '.' 0.1 ' ' 1
Residual standard error: 0.9574 on 142 degrees of freedom
(10 observations deleted due to missingness)
Multiple R-squared: 0.2521,     Adjusted R-squared: 0.2363
F-statistic: 15.95 on 3 and 142 DF, p-value: 5.414e-09
```

Although it is more difficult to see in this case due to variance being partitioned slightly differently in the two approaches, we can focus on the interaction effect. The p value for the interaction is identical for the two approaches $p = .640$. The effect is tested with an F value for the ANOVA ($F = .22$) and a t value for the regression ($t = -.469$); however if you square the t value you obtain the F value $(-.469)^2 = .22$

ANOVA OR REGRESSION?

Now that the equivalence of regression and ANOVA has been demonstrated, where do we go from here? Is ANOVA even necessary? As may have already been suspected, regression is the parent model. All ANOVA models are essentially special cases of regression and can be run using regression models. So why do we need ANOVA at all? Seems redundant. This is a valid perspective, and mathematically speaking it is correct. However, just because regression and ANOVA are mathematically equivalent, this does not mean that ANOVA doesn't have a place in the statistical toolbox. Both ANOVA and regression use the same mathematical model to emphasize different viewpoints and thus each has situations that they are best at handling. Here is a short list of considerations you may wish to make as you are pondering which approach is best for your analysis.

1. Both regression and ANOVA require a continuous dependent variable; however, ANOVA works best when there are multiple categorical variables (factors) to be used as independent variables. Regression works best when there are multiple continuous variables (covariates) to be used as independent variables. Yes, both ANOVA and regression can use both continuous and categorical independent variables but if you find yourself using one factor and five covariates, you may want to use regression instead of ANOVA. If you find yourself using one covariate and three factors, you may want to use ANOVA.

2. As seen above, dummy variables in regression can be used to look at significant differences between groups the same way as is done in ANOVA. However, due to multicollinearity issues, only $k - 1$ of the groups may be compared at a time in a regression. If all groups are to be compared using ANOVA, the model would need to be run multiple times with different dummy coding schemes. This is not a limitation of ANOVA models. Therefore, if the focus of analysis is to determine significant differences between groups, ANOVA may be better suited to the purpose.

3. Regression and ANOVA answer different research questions. Although both models can address the question of variance in the outcome explained by the independent (explanatory) variables in the model, regression's main focus is identifying significant predictors of the outcome. What are the strongest predictors of the outcome and in what way do they relate? ANOVA's main focus is on group differences. Are there significant differences in the outcome between independent variable groups?

4. Due to the differences in emphasis, ANOVA methods can have slightly different (often more strict) assumptions than regression. Take for example, analysis of covariance (ANCOVA). ANCOVA introduces covariates (continuous independent variables) into an ANOVA model. It does so in a very deliberate way that assumes that the covariate has an equal impact at all levels of the independent variable. This is the assumption of homogeneity of regression slopes. If this assumption of ANCOVA is violated, there is little that can be done because this is an actual effect, not a characteristic that can be manipulated (like normality). Instead, the researcher could choose to look at this from a regression perspective. This assumption is not

made in regression and regression methodology would allow the differential impact of the covariate to be easily assessed (using interaction terms).

5. Yes, ANOVA can be done using regression with dummy variables to represent the factors in the model. But consider how unwieldy this process would become with a categorical variable with five levels. Multiple categorical variables with many levels? Interactions between these multilevel categorical variables? Although mathematically possible, this situation is simply easier to handle using ANOVA which is built to handle categorical variables.

Bottom line? In many situations either regression or ANOVA could be used. It often comes down to which better fits the research purpose (emphasis on group differences or emphasis on predictors) and which is easiest! It is useful to understand the equivalence of these methods, however, because it puts both methods into better perspective. And many individuals will start noticing the similarities between these methods well before knowing that they are mathematically equivalent. As a professor, I get asked these questions all the time. 'Is the R squared in regression similar to the eta squared in ANOVA?' Why yes. Yes it is.

Model Comparisons and Hierarchical Regression

THE PURPOSE OF THIS chapter is to discuss methodologies for comparing regression models. To motivate our discussion, we will again turn to an example from the *Mask* dataset used in the previous chapters. The *Mask* dataset is a sample of college students from a midsized midwestern university which was collected during the COVID-19 pandemic. The main research questions of interest revolved around mask wearing and whether mask wearing and attitudes toward mask wearing were related to various personality characteristics and hearing/vision impairment.

WHY COMPARE MODELS?

Before talking about methodologies for comparing models, it is important to understand different motivations for performing model comparisons. There are a number of reasons why the researcher may want to perform comparisons between regression models. These reasons are generally based on the theoretical rationale for the multiple models defined. Below you will find a list of common reasons for model comparisons.

1. *Controlling for demographic/confounding variables.* In many research projects, the researcher may have reason to control specific variables which, although suspected to be related to the outcome of interest, are generally not of great interest to the researcher. Often this includes relevant demographic variables or potential confounding variables

DOI: 10.1201/9780429295843-7

that have already been well established as important by previous literature. *For example, a researcher may wish to test whether worry about catching COVID-19 can be predicted by sleep difficulty, worry over family safety, and mask wearing after controlling for the age and gender of the individual.*

2. *Theoretical priority order.* There are also many times when a researcher hypothesizes certain variables to be of greater importance than other variables when accounting for variability in the outcome. This priority is generally based on specific research questions of interest, or the strength of predictors as identified in previous literature. *For example, a researcher may wish to predict mask frustration from personality characteristics and vision impairment giving priority to vision impairment due to its importance in the literature.*

3. *Selecting predictors for parsimony.* Often a number of relevant predictors are identified and researchers are interested in finding the best set of predictors to maximize variance explained without using unnecessary predictors. This predictor selection is not based on theoretical rationale or literature review but rather by computer algorithm. *For example the researcher may want the best most parsimonious model predicting Mask Attitude considering all measured variables as potential variables in the model.*

4. *Competing or similar models.* There are often times when a researcher chooses to run a number of similar regression models differing in only certain aspects. This may include splitting a sample and comparing a regression model across groups or timepoints or comparing models using different but similar predictors. *For example, the researcher may run a regression predicting mask frustration but choose to run the model separately for males and females.*

Each of these scenarios would be used for very different types of research questions. However, the common factor across each is that multiple regression models would need to be run and systematically compared in order to fully answer the research question. In the next section we will talk about methods for comparing both nested regression models (this would address points 1 and 2 and 3) and non-nested regression models (this would address point 4). It is important to differentiate between nested and non-nested models as the statistical methodology used for each situation is necessarily different.

WHAT DOES IT MEAN FOR MODELS TO BE NESTED?

Nested regression models occur when each model compared can be obtained by simply adding or deleting predictors from the other compared models. Or another way to phrase is that each smaller model is contained in the larger models. This may be best illustrated with an example. Consider the set of regression models below. Mask.model1 is nested within mask.model2. The entire predictor structure of mask.model1 (gender and age) is contained within mask.model2. Or another way to think about this is that we could obtain mask.model2 by simply adding predictors to mask.model1.

$$\text{Mask.model1: } \widehat{MaskAtt} = a + b_1 \left(gender \right) + b_2 \left(age \right) \tag{7.1}$$

$$\text{Mask.model2: } \widehat{MaskAtt} = a + b_1 \left(gender \right) + b_2 \left(age \right) + b_3 \left(familyworry \right)$$
$$+ b_4 \left(sleepdist \right) \tag{7.2}$$

Similarly, consider the sets of models below. Mask.model3 and Mask.model4 are non-nested models. You cannot obtain mask.model4 from mask.model3 (or vice versa) by adding or subtracting predictors because these models have different predictor sets. Similarly, Mask.model5 and Mask.model6 are also non-nested models. You cannot obtain one model from the other by adding or subtracting predictors because they have different outcome variables.

$$\text{Mask.model3: } \widehat{MaskAtt} = a + b_1 \left(gender \right) + b_2 \left(age \right) + b_3 \left(familyworry \right) \tag{7.3}$$

$$\text{Mask.model4: } \widehat{MaskAtt} = a + b_1 \left(gender \right) + b_2 \left(age \right) + b_3 \left(sleepdist \right) \tag{7.4}$$

$$\text{Mask.model5: } \widehat{MaskAtt} = a + b_1 \left(gender \right) + b_2 \left(age \right) \tag{7.5}$$

$$\text{Mask.model6: } \widehat{MaskEff} = a + b_1 \left(gender \right) + b_2 \left(age \right) \tag{7.6}$$

MODEL COMPARISONS FOR NESTED AND NON-NESTED MODELS

Now that you can recognize whether models are nested or non-nested, the discussion can return to model comparisons. It is important to know if your model comparison is nested or non-nested because model comparison procedures are different for nested and non-nested models.

Comparisons of Non-Nested Models

When models are non-nested, a general model comparison statistic is necessary in order to compare models. Most commonly this is done with the Akaike information criterion (AIC) and/or the Bayesian information criterion (BIC). The AIC and BIC are very general measures of model fit that can be used to compare the fit of models of any type, nested or non-nested, regression, structural equation model, hierarchical linear model, etc. AIC and BIC work by measuring the lack of fit for each model and penalize this value for the number of parameters (predictors) included in the model. Thus, arguing for parsimony: is improved fit worth the number of predictors used?

Smaller AIC and BIC are indicative of better model fit, or more accurately, less 'lack of fit'. Being very general, AIC and BIC have no associated significance tests so interpretation needs to be fairly loose. For example, if the AIC and BIC for two models only differ by one or two points, it may be best to say the model fit is essentially equal. It is also important to note that AIC and BIC are non-interpretable for just a single model. They only have meaning in comparison with other AIC and BIC. Thus, you CAN say 'comparing the AICs, Model 1 fits better than Model 2'. You CANNOT say 'Based on the AIC, Model 1 fits well'.

R Example of Non-Nested Model Comparison

EXAMPLE: ARE AGE, GENDER AND COVID-RELATED FAMILY WORRY BETTER PREDICTORS OF MASK ATTITUDE OR MASK-RELATED FRUSTRATION? USING MASK DATASET.

Clearly this is a question that could be answered by multiple indicators: we could look at the coefficients for each model to see which predictors are significant and strongest; we could look at the R squared or standard error of the estimate. But most importantly, we can use the AIC and BIC comparison statistics to compare the fit of the two models to see which is a better fit model.

We can begin by fitting a regression for each model.

```
MaskAtt <- lm(MaskAtt~Age+GenderIdentity+CSS _ familyworry,
Mask)
summary(MaskAtt)
```

```
Call:
lm(formula = MaskAtt ~ Age + GenderIdentity + CSS _ familyworry,
data = Mask)
Residuals:
Min    1Q  Median    3Q    Max
-13.3016 -2.2692  0.5504  2.6960  8.4472
Coefficients:
    Estimate Std. Error t value Pr(>|t|)
(Intercept)    12.65524  1.99174   6.354 2.38e-09 ***
Age           -0.07401   0.13448  -0.550   0.583
GenderIdentity 1.19361   0.95515   1.250   0.213
CSS _ familyworry 1.63879 0.25422   6.446 1.48e-09 ***
---
Signif. codes: 0 '***' 0.001 '**' 0.01 '*' 0.05 '.' 0.1 ' ' 1
Residual standard error: 4.426 on 150 degrees of freedom
(2 observations deleted due to missingness)
Multiple R-squared: 0.2373, Adjusted R-squared: 0.222
F-statistic: 15.56 on 3 and 150 DF, p-value: 7.336e-09
```

```
MaskFrust    <-lm(NASA _ 1 _ 4~Age+GenderIdentity+CSS _ familywo
rry, Mask)
summary(MaskFrust)
```

```
Call:
lm(formula = NASA _ 1 _ 4 ~ Age + GenderIdentity + CSS _ familyworry,
data = Mask)
Residuals:
Min    1Q  Median    3Q    Max
-9.480 -4.628 -2.689 4.599 16.098
Coefficients:
    Estimate Std. Error t value Pr(>|t|)
(Intercept)    8.86066  3.08696   2.870 0.004728 **
Age           -0.05825   0.20586  -0.283 0.777635
GenderIdentity 0.36788   1.48400   0.248 0.804571
CSS _ familyworry -1.42595 0.39430  -3.616 0.000414 ***
---
Signif. codes: 0 '***' 0.001 '**' 0.01 '*' 0.05 '.' 0.1 ' ' 1
Residual standard error: 6.738 on 142 degrees of freedom
(10 observations deleted due to missingness)
Multiple R-squared: 0.0866, Adjusted R-squared: 0.06731
F-statistic: 4.488 on 3 and 142 DF, p-value: 0.00484
```

Looking at the output for each model, we can see that Age, Gender, and Family Worry account for 23% of the variance in Mask Attitude but only 8% of the variance in Mask Frustration. For both models, Family Worry was the only significant predictor.

For additional information, we can use the `AIC()` and `BIC()` functions from the base R package.

```
AIC(MaskAtt)
AIC(MaskFrust)
BIC(MaskAtt)
BIC(MaskFrust)

>AIC(MaskAtt)
[1]  901.1237
> BIC(MaskAtt)
[1]  916.3085
> AIC(MaskEff)
[1]  915.3213
> BIC(MaskEff)
[1]  930.1355
```

Comparing AIC and BIC statistics we can see that both the AIC and BIC for Mask Attitude are smaller than the AIC and BIC for Mask Effort indicating Mask Attitude is the better fit.

COMPARISONS OF NESTED MODELS

The most common model comparison statistic for nested regression models is the R^2 change statistic. R^2 change is exactly what it sounds like: how much the model R^2 changed or increased (R^2 model 2 – R^2 model1) as predictors were added to the regression models (creating nested regressions).

It is not enough to just demonstrate that the R^2 increases as the model becomes more complex. Mathematically speaking, variance explained will always increase as variables are added to the model. Thus, the question is not if the variance increased with the additional predictors, but if the variance increased *enough* with the additional predictors, to justify the addition of those predictors. Consider this again as an argument for parsimony. We don't want the extra model complexity if it only barely increases the variance explained. Thus, we generally don't just look at how much the R squared changed but if the change in R squared is significant from one model to the next. The significance of R^2 change can be tested using an F statistic computed based on the sums of squares (SS), degrees of freedom (df), and mean square residual (MS residual) of the full model (model with more predictors) and reduced model (model with fewer predictors).

$$F = \frac{\dfrac{SSreg\,(full) - SSreg\,(reduced)}{dfreg\,(full) - dfreg\,(reduced)}}{MSresidual\,(full)} \qquad (7.7)$$

Types of Nested Model Comparison

Sequential regression. Sequential regression refers to nested regression model comparison where the choice of models to be compared is determined by the researcher based on theory or previous literature. These models are purposefully and strategically tailored to answer specific research questions, represent theoretical variable priority, or address methodological issues (such as controlling for potential confounding variables). *The most defining characteristics of these models are that they are nested models and the choice of variable entry/priority is researcher driven.*

Stepwise regression. Stepwise regression refers to nested regression model comparisons where the choice of models to be compared is determined by computer algorithm. There are a variety of common computer algorithms for stepwise regression but all share the characteristic that predictor entry/ deletion choice is determined by the algorithm, not by the researcher.

Forward regression. When the researcher selects a forward regression algorithm, the computer will select and enter predictors one at a time. Predictors will be selected such that the predictor which will explain the most variance will be entered first. Then the model and predictors will be reevaluated and the next predictor will be selected which provides the largest increase in variance explained. The model and predictors will be reevaluated again. This process will continue to occur until all predictors have been entered or no predictors are left which significantly increase the explained variance.

Backward deletion. When the researcher selects a backward deletion algorithm, the computer will begin with all identified predictors entered into the model. Then predictors will be deleted from the model one at a time beginning with the predictor which accounts for the least variance. With each deletion the model will be reevaluated, the next least important predictor will be deleted, and so on, until there are no unnecessary predictors left: any deletion would result in a significant reduction in variance explained. A good way to think about a backward deletion model is 'trimming the fat'.

Stepwise regression. Stepwise regression is a combination of forward selection and backward deletion. Stepwise regression generally begins with no predictors and adds predictors one at a time (like forward regression). At each step, however, the computer will decide the best course of action to maximize variance explained. This may mean variables are added or deleted to arrive at the best fitting model that maximizes variance explained with the fewest unnecessary predictors.

R Example of Sequential Regression

EXAMPLE: HOW WELL DOES COVID-RELATED FAMILY WORRY PREDICT MASK ATTITUDE AFTER TAKING INTO ACCOUNT DEMOGRAPHIC CHARACTERISTICS (AGE AND GENDER) OF THE INDIVIDUAL? USING MASK DATASET.

We will begin by running a model containing the demographic variables only (CovidAnxDemog) and a model that adds the third predictor, Family Worry, to the model (CovidAnx).

```
CovidAnxDemog <-lm(MaskAtt~Age+Gender, Mask)
summary(CovidAnxDemog)

Call:
lm(formula = MaskAtt ~ Age + Gender, data = Mask)
Residuals:
Min   1Q Median   3Q   Max
-12.588 -2.588  1.412  3.444  7.355
Coefficients:
 Estimate Std. Error t value Pr(>|t|)
(Intercept) 14.83175  2.22266  6.673 4.47e-10 ***
Age      -0.01623  0.14997 -0.108  0.9140
Gender    1.89457  1.08981  1.738  0.0842.
---
Signif. codes: 0 '***' 0.001 '**' 0.01 '*' 0.05 '.' 0.1 ' ' 1
Residual standard error: 4.995 on 151 degrees of freedom
(2 observations deleted due to missingness)
Multiple R-squared: 0.02219,Adjusted R-squared: 0.009242
F-statistic: 1.714 on 2 and 151 DF, p-value: 0.1837


CovidAnx <-  lm(MaskAtt~Age+GenderIdentity+CSS _ familyworry,
Mask)
summary(CovidAnx)
```

```
Call:
lm(formula = MaskAtt ~ Age + GenderIdentity + CSS _ familyworry,
data = Mask)
Residuals:
Min    1Q  Median    3Q   Max
-13.3016 -2.2692  0.5504  2.6960  8.4472
Coefficients:
    Estimate Std. Error t value Pr(>|t|)
(Intercept)   12.65524  1.99174  6.354 2.38e-09 ***
Age          -0.07401  0.13448 -0.550  0.583
GenderIdentity 1.19361  0.95515  1.250  0.213
CSS _ familyworry 1.63879  0.25422  6.446 1.48e-09 ***
---
Signif. codes: 0 '***' 0.001 '**' 0.01 '*' 0.05 '.' 0.1 ' ' 1
Residual standard error: 4.426 on 150 degrees of freedom
(2 observations deleted due to missingness)
Multiple R-squared: 0.2373,Adjusted R-squared: 0.222
F-statistic: 15.56 on 3 and 150 DF, p-value: 7.336e-09
```

The demographics only model accounts for only 2% of the variance in mask attitude. Once covid-related family worry is added to the model the model accounts for 23% of the variance in mask attitude. This would be an R^2 change of .21.

To see if this is a significant increase in variance explained, we can use the anova() command to compare the models.

```
anova(CovidAnxDemog, CovidAnx)

Analysis of Variance Table
Model 1: MaskAtt ~ Age + Gender
Model 2: MaskAtt ~ Age + GenderIdentity + CSS _ familyworry
Res.Df  RSS Df Sum of Sq   F  Pr(>F)
1 151 3767.0
2 150 2938.3 1  828.67 42.303 1.097e-09 ***
---
Signif. codes: 0 '***' 0.001 '**' 0.01 '*' 0.05 '.' 0.1 ' ' 1
```

Using this code, the demographics only model is 'Model 1' and the full model is 'Model 2'. The output shows the difference in the residual variance, degrees of freedom, sum of squares, and provides the F test for the R squared change between the two models. The F for the change in R squared is 42.301, p < .001 indicating that the addition of covid-related family worry significantly increases the variance explained above the demographics alone.

R Examples of Stepwise Regression

EXAMPLE: WHAT IS THE MOST PARSIMONIOUS MODEL PREDICTING COVID-19 ANXIETY FROM ANXIETY OVER FAMILY MEMBERS, REPORTED SLEEP DISTURBANCE, AGE, AND GENDER? USING MASK DATASET.

The step() function provides a very simple process for running stepwise regression models.

First begin by running an intercept only model and a full (all predictors) model.

```
interceptonly <- lm(CSS _ 1 _ 1~1, Mask)
allpredictors <- lm(CSS _ 1 _ 1 ~ Gender+Age+CSS _ family
worry+CSS _ sleep, Mask)
```

Forward Regression Solution

The model with no predictors (interceptonly) is the starting point for the forward regression. The full model with all predictors (allpredictors) is the scope of the model. This tells the forward regression what predictors it has to select from. Then the direction option selects the type of stepwise solution (forward, backward, or both).

```
mask.model2for<-step(interceptonly, scope=
formula(allpredictors), direction="forward")
summary(mask.model2for)

mask.model2for$anova
Coefficients:
    Estimate Std. Error t value Pr(>|t|)
(Intercept)   0.43727  0.17982  2.432 0.01619 *
CSS _ familyworry 0.47276  0.06083  7.772 1.07e-12 ***
CSS _ sleep    0.29229  0.11151  2.621 0.00966 **
Age       0.04963  0.02996  1.657 0.09960.
---
Signif. codes: 0 '***' 0.001 '**' 0.01 '*' 0.05 '.' 0.1 ' ' 1
Residual standard error: 1.016 on 152 degrees of freedom
Multiple R-squared: 0.3901,Adjusted R-squared: 0.378
F-statistic: 32.4 on 3 and 152 DF, p-value: 2.989e-16

mask.model2for$anova

  Step Df Deviance Resid. Df Resid. Dev    AIC
1          NA    NA    155  257.1474 79.967801
2 + CSS _ familyworry -1 87.759445  154  169.3880 16.844398
3    + CSS _ sleep -1 9.716480   153  159.6715 9.628972
4      + Age -1 2.832792   152  156.8387 8.836472
```

The forward regression selected three predictors with familyworry entering first, followed by sleep and then age. Using the anova() call, the AIC can be seen for each step of the forward procedure. Notice the AIC getting smaller and smaller indicating better model fit with each additional predictor.

BACKWARD SOLUTION

To run a backward model we use the all predictors model as a starting point. Then if we set the direction to backward, R will use the backward stepwise criteria to select a best fit model.

```
mask.model2back <- step(allpredictors, direction=
"backward")
summary(mask.model2back)
mask.model2back$anova
```

```
Coefficients:
    Estimate Std. Error t value Pr(>|t|)
(Intercept)   0.43727  0.17982  2.432 0.01619 *
Age           0.04963  0.02996  1.657 0.09960.
CSS _ familyworry 0.47276 0.06083 7.772 1.07e-12 ***
CSS _ sleep   0.29229  0.11151  2.621 0.00966 **
---
Signif. codes: 0 '***' 0.001 '**' 0.01 '*' 0.05 '.' 0.1 ' ' 1
Residual standard error: 1.016 on 152 degrees of freedom
Multiple R-squared: 0.3901, Adjusted R-squared: 0.378
F-statistic: 32.4 on 3 and 152 DF, p-value: 2.989e-16
> mask.model2back$anova
Step Df Deviance Resid. Df Resid. Dev     AIC
1      NA    NA      151   156.6544 10.653079
2 - Gender 1 0.1842706   152   156.8387 8.836472
```

The backward elimination criteria selected a model including age, family-worry, and sleep as the best fit model. Using the anova() call we can see that the best solution was arrived at in one step simply by deleting gender in step 1. We can see that this is exactly the same solution arrived at by the forward regression (AIC 8.836 also identical). This solution was just obtained through a different procedure.

STEPWISE SOLUTION

To run a stepwise model we will again use the intercept only model as a starting point and all predictors model as the scope. Then if we set the direction to both, R will use the stepwise criteria to select a best fit model. (The both referring to allowing both entry and deletion of predictors at each step.)

```
mask.model2step  <-     step(interceptonly,     scope=formula
(allpredictors), direction="both")
```

```
summary(mask.model2step)
mask.model2step$anova

Coefficients:
    Estimate Std. Error t value Pr(>|t|)
(Intercept)     0.43727  0.17982   2.432 0.01619 *
CSS _ familyworry 0.47276  0.06083   7.772 1.07e-12 ***
CSS _ sleep     0.29229  0.11151   2.621 0.00966 **
Age          0.04963  0.02996   1.657 0.09960.
---
Signif. codes: 0 '***' 0.001 '**' 0.01 '*' 0.05 '.' 0.1 ' ' 1
Residual standard error: 1.016 on 152 degrees of freedom
Multiple R-squared: 0.3901, Adjusted R-squared: 0.378

F-statistic: 32.4 on 3 and 152 DF, p-value: 2.989e-16
> mask.model2step$anova
   Step Df Deviance Resid. Df Resid. Dev    AIC
1          NA    NA     155   257.1474 79.967801
2 + CSS _ familyworry -1 87.759445   154   169.3880 16.844398
3    + CSS _ sleep -1 9.716480    153   159.6715 9.628972
4       + Age -1 2.832792    152   156.8387 8.836472
```

In this case, the stepwise criteria selected the same three predictors as the forward model with familyworry entering first, followed by sleep and then age. Deletion of variables was not deemed necessary in this circumstance.

CHAPTER SUMMARY

In summary, this chapter provides methodology for using multiple regression to answer specific types of research questions where comparison of multiple models is of relevance. As is described, there are many different reasons why a comparison of models would be of relevance and depending on the type of model comparison to be done, different statistics and methodologies may be necessary. When considering methodologies for model comparisons the researcher should always consider the questions:

1. What is the purpose of my comparison?

2. Is there a theoretically or conceptually driven order I want to use to drive my comparisons or is there reason to just find the most parsimonious model that fits?

3. Should I be the one making my modeling decisions or should I let the computer decide?

4. Will this produce nested or non-nested models and as such, what is the best way to assess the best fit model?

Consideration of these questions should help lead the researcher to make good decisions regarding what is the best and most ethical way to proceed.

CHAPTER 7: END OF CHAPTER EXERCISES

Using dataset chapter7ex

1a. Run a multiple regression model predicting attention from screen time, age, and boy.

1b. Run a multiple regression model predicting attention from screen time, age, boy, and the interaction of screen time and age.

1c. Compare the models from 1a and 1b. Does the addition of the interaction significantly increase the variance explained above and beyond the main effects?

2. Return to the model in 1a.

  a. Use these predictors to run a forward regression. What was identified as the best fit model?

  b. Use these predictors to run a backward regression. What was identified as the best fit model?

3a. Run a model predicting attention from screen time and age.

3b. Run a model predicting attention from screen time and boy.

3c. Compare AIC and BIC statistics for these two models. Which is the best fit?

Regression Extensions 1

Moderation/Mediation and Regression Discontinuity

*T*AKING STOCK: AT THIS point this text has covered interpretation and R code for running models using the basic OLS regression framework. This chapter will now combine procedures presented in the previous chapters in order to answer some richer types of research questions. For example, interactions can be used to answer research questions involving moderating variables and treatment/control research. Hierarchical regression can be used to answer research questions involving mediating relationships. This chapter will briefly address each of these types of research question and provide examples of how regression methodology can be used in these research situations.

EXTENSION 1: MODERATION

Motivating Example

You are a researcher interested in determining the strongest motivating factors for high school students learning algebra. You would like to test whether the teacher's enthusiasm toward algebra moderates the relationship between students' attitude toward algebra and their subsequent algebra performance.

What do we mean by 'moderation'? Moderation refers to situations when a third variable changes the relationship between two other variables. Sometimes when interpreting moderating relationships you

DOI: 10.1201/9780429295843-8

will hear phrasing like the 'relationship was *buffered by*', 'the relationship was *enhanced by*', or 'the relationship *depended on*'. This phrasing should feel familiar as all of it describes situations where the effect of one variable on another is different depending on the level of a third variable. In other words, an interaction. Knowing that moderation refers to interactions, I prefer to think about moderation as a specific phrasing of a research question that helps give interpretational direction to an interaction effect.

R EXAMPLE: MODERATION

Motivating example: When looking at the relationship between a high school student's attitude toward algebra and their performance in algebra, does the enthusiasm level of the teacher moderate this relationship?

(Using Algebra data).

As moderation is just a question of interaction, this research question can be translated to: can we predict a student's performance in algebra from their attitude toward algebra, the enthusiasm level of their teacher, and the interaction between attitude toward algebra and teacher enthusiasm?

```
Alg _ Mod <-  lm(Alg _ Perf~AlgebraAtt+Inst _ Enthus+exp+Algebr
aAtt*Inst _ Enthus, AlgebraData)
summary(Alg _ Mod)

> summary(Alg _ Mod)
Call:
lm(formula = Alg _ Perf ~ AlgebraAtt + Inst _ Enthus + AlgebraAtt *
  Inst _ Enthus, data = AlgebraData)
Residuals:
Min   1Q Median   3Q  Max
-45596 -11893  713 12048 45345
Coefficients:
             Estimate Std. Error t value Pr(>|t|)
(Intercept)      1584.861 26081.591  0.061  0.9516
AlgebraAtt        -54.792  166.983 -0.328  0.7430
Inst _ Enthus    -154.069  867.080 -0.178  0.8591
AlgebraAtt:Inst _ Enthus 11.787   5.577  2.113  0.0353 *
---
Signif. codes: 0 '***' 0.001 '**' 0.01 '*' 0.05 '.' 0.1 ' ' 1
Residual standard error: 17380 on 346 degrees of freedom
Multiple R-squared: 0.4679, Adjusted R-squared: 0.4633
F-statistic: 101.4 on 3 and 346 DF, p-value: < 2.2e-16
>
```

Results show that the model is significant accounting for 46% of the variance in algebra performance. Although neither main effect is significant, the interaction of attitude toward algebra and instructor enthusiasm is significant initially suggesting that yes, the relationship between algebra attitude and algebra performance is moderated by instructor enthusiasm.

```
library(interactions)
johnson _ neyman(Algebradata, AlgebraAtt, Inst _ Enthus, vmat
= NULL, alpha = 0.05,
     plot = TRUE, title = "Johnson-Neyman plot")
```

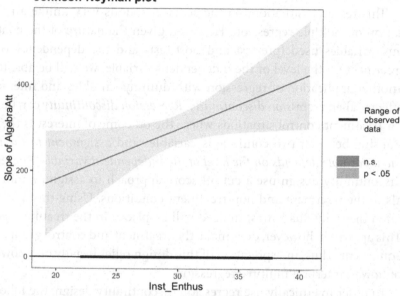

Johnson-Neyman plot

```
JOHNSON-NEYMAN INTERVAL
When Inst _ Enthus is OUTSIDE the interval [-332.55, 16.94],
the slope of
AlgebraAtt is p<.05.
Note: The range of observed values of Inst _ Enthus is [22.11,
37.68]
```

To interpret this moderated relationship further we can look at the Johnson-Neyman output. Looking at the plot it is clear that the slope for algebra attitude increases as instructor enthusiasm increases. Thus, instructor enthusiasm has an enhancing effect on this relationship: as instructor enthusiasm increases, algebra attitude becomes an increasingly stronger positive predictor of algebra performance.

EXTENSION 2: REGRESSION DISCONTINUITY

Motivating Example

You have a sample of elementary school students who are pre-tested in reading at the beginning of the fourth grade and then again at the end of the fourth grade to see how their reading achievement has improved. Looking at the pre-test scores, there are a number of students struggling to read at fourth-grade level so you decide to implement a special reading program for all students who score below grade level. You want to determine if the reading program was successful.

This research situation, on the surface, sounds very unlike an application of multiple regression. However, given the nature of the continuous variables used (pre-test and post-test) and the dependency of the treatment on the level of the independent variable, we will be able to use another application of regression with dummy variables and interaction effects called *regression discontinuity*. *Regression discontinuity* can be used for treatment/control situations where the outcome of interest is the relationship between two continuous variables and *assignment to the treatment variable depends on the level of the independent variable*. Regression discontinuity design use a cut off score approach to assigning individuals to the treatment and non-treatment conditions. Using this approach, often the individuals most in need will be placed in the treatment group. This approach, however, does make the treatment and control groups nonequivalent. Thus, interpretation of this design relies heavily on knowledge of how to interpret multiple regression.

In order to ethically use regression discontinuity design, the following requirements must be in place.

1. The cut off score must be strictly followed when determining treatment/non-treatment groups.

2. The treatment must be uniformly delivered to all individuals in the treatment group.

3. The relationship of interest must be between continuous variables.

4. There must be a sufficient number for regression in the treatment and non-treatment groups and there must be sufficient variability in each group.

5. The relationship of interest must be linear and the outcome must be described by a polynomial function able to be modeled by a linear regression line.

Interpreting Treatment Effects in Regression Discontinuity Design

In a Regression Discontinuity Design, discontinuity can be evidence of two different types of differences: intercept differences and slope differences.

Intercept differences (as evidenced by a significant dummy variable for the treatment) can be taken as evidence of an effect of treatment on the average performance of the individuals. Visually this will take the form of a discontinuity in the regression line where the intercept of the line will jump at the point where the control group switches to the treatment group.

Slope differences (as evidenced by a significant interaction between the dummy variable for treatment and the post-test) can be taken as evidence of a treatment effect on the relationship between the pre-test and the post-test. Slope differences can be present regardless of intercept differences. So it is possible to have a treatment effect where just the relationship between pre- and post-test changes depending on the treatment, or it is possible for the intercepts to be as well as the relationship between the pre- and the post-test.

*A Note on the Terminology

When describing regression discontinuity designs, it is often easiest to talk about pre-test/post-test examples. However, regression discontinuity can be applied to any research situation where the relationship between two variables may be impacted by the cut off score on one of the variables. Other examples may include examining the relationship between household income and access to food (access to support services based on cut off point) or relationship between high school and college GPA (cut off point representing access to scholarships).

> *Motivating example: the following data look at reading scores for fourth grade students. A reading pre-test was given and students achieving below the median score were given a reading intervention. There is a strong relationship between reading pre-test and reading post-test, so if the treatment given to the at-risk readers has an impact, there should be a discontinuity to the regression line.*

Regression Discontinuity Example 1. No Discontinuity (No Treatment Effect)

First, we will test a simple discontinuity model with the only difference in the intercepts of the lines. This is just an application of using a dummy variable for treatment. This example will illustrate what a lack of discontinuity looks like in both the output and visually.

```
nodiscont<-lm(readingpost~readingpre+treatment,
NOdiscontdata)
summary(nodiscont)
NOdiscontdata$predno<-predict(nodiscont)


Call:
lm(formula = readingpost ~ readingpre + treatment,   data =
Discontdata)
Residuals:
 Min    1Q  Median    3Q   Max
-17.0864 -4.0441  0.7257  4.1869 14.2403
Coefficients:
      Estimate Std. Error t value Pr(>|t|)
(Intercept) 173.1350    1.3182 131.343 < 2e-16 ***
readingpre  2.2742    0.7673  2.964 0.00354 **
treatment   2.8235    1.7070  1.654 0.10025
---
Signif. codes: 0 '***' 0.001 '**' 0.01 '*' 0.05 '.' 0.1 ' ' 1
Residual standard error: 6.271 on 147 degrees of freedom
Multiple R-squared: 0.05999, Adjusted R-squared: 0.0472
F-statistic: 4.691 on 2 and 147 DF, p-value: 0.0106
```

This model is significant but only accounts for 6% of the variance in post-test reading. The dummy variable for treatment is not significant (p= .1) indicating no differential effect of the treatment. Looking at the plot below we can also see a lack of discontinuity to the regression line. Plotting this type of model can be done by making a basic scatterplot, then creating two different lines, one line for the treatment = 0 group and one line for the treatment = 1 group. In this plot the treatment line was thickened to make it stand out.

```
plot(NOdiscontdata$readingpre,NOdiscontdata$readingpost)
with(subset(NOdiscontdata, treatment==0),lines(readingpre,
predno)) with(subset(NOdiscontdata,
treatment==1),lines(readingpre, predno,
                     type = "l", lwd = "3"))
```

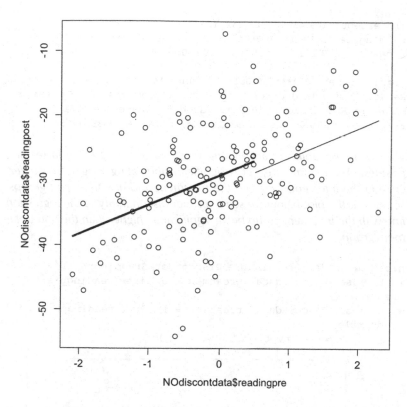

Regression Discontinuity Example 2: No Discontinuity (Treatment Effect)

Now, we will test a simple discontinuity model again but this example will illustrate what a discontinuity (treatment effect) will look like.

```
discontsimple<-(lm(readingpost~readingpre+treatment,
Discontdata))
summary(discontsimple)
Discontdata$predsimple<-predict(discontsimple)

Call:
lm(formula = readingpost ~ readingpre + treatment,  data =
Discontdata)
Residuals:
 Min    1Q  Median    3Q    Max
-21.7349 -5.1382  0.1765  5.0709 19.3404
Coefficients:
    Estimate Std. Error t value Pr(>|t|)
```

```
(Intercept) 169.4525    1.3851 122.339 < 2e-16 ***
readingpre  5.1289    0.8612   5.956 1.83e-08 ***
treatment   7.1517    1.9841   3.604 0.000428 ***
---
Signif. codes: 0 '***' 0.001 '**' 0.01 '*' 0.05 '.' 0.1 ' ' 1
Residual standard error: 7.276 on 147 degrees of freedom
Multiple R-squared: 0.2059, Adjusted R-squared: 0.1951
F-statistic: 19.06 on 2 and 147 DF, p-value: 4.358e-08
```

This model is significant accounting for 20% of the variance in reading post-test. The treatment effect is significant indicating a significant difference in the intercepts between the treatment and non-treatment groups. Looking at the plot below we see a distinct discontinuity to the regression line with the treatment group being significantly higher than the non-treatment group.

```
plot(Discontdata$readingpre,Discontdata$readingpost)
with(subset(Discontdata, treatment==0),lines(readingpre,
predsimple))
with(subset(Discontdata, treatment==1),lines(readingpre,
predsimple,
                    type = "l",  lwd = "3"))
```

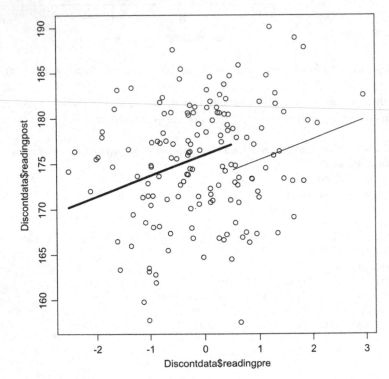

Regression Discontinuity Example 3: Discontinuity with Varying Slopes

Now, we will test a discontinuity model where the slopes are allowed to vary. In this model we will see both a treatment effect on the intercept as well as differences in the slope.

```
discontflex      <-lm(readingpost~readingpre+treatment+reading
pre*treatment, Discontdataflex)
summary(lm.beta(discontflex))
Discontdataflex$pred <- predict(discontflex)


Call:
lm(formula = readingpost ~ readingpre + treatment + readingpre *
  treatment, data = Discontdataflex)
Residuals:
 Min    1Q  Median    3Q    Max
-18.3097 -3.5178  0.1214  4.1461 18.7119
Coefficients:
         Estimate Standardized Std. Error t value Pr(>|t|)
(Intercept)    168.0583    0.0000   2.5505 65.893 <2e-16 ***
readingpre       4.8522    0.5305   1.9720  2.461  0.0150 *
treatment        5.8862    0.2708   2.6613  2.212  0.0285 *
readingpre:treatment 4.8693 0.3616   2.1299  2.286  0.0237 *
---
Signif. codes: 0 '***' 0.001 '**' 0.01 '*' 0.05 '.' 0.1 ' ' 1
Residual standard error: 6.578 on 146 degrees of freedom
Multiple R-squared: 0.5367,Adjusted R-squared: 0.5272
F-statistic: 56.38 on 3 and 146 DF, p-value: < 2.2e-16
```

Here we see both main effects of reading pre-test and treatment are significant as well as the interaction between pre-test and treatment. The significance of the treatment variable means that we do have a regression discontinuity with the treatment group having a significantly higher intercept than the non-treatment group. The significance of the interaction effect indicates that the slopes for the treatment and non-treatment group also differ. This is visually very clear in the plot below where we see a considerably stronger positive slope for the treatment group than for the non-treatment group. Thus, the treatment group had both significantly higher average reading post-tests as well as a stronger relationship between pre-test and post-test.

```
plot(Discontdataflex$readingpre,Discontdataflex$readingpost)
with(subset(Discontdataflex, treatment==0),lines(readingpre, pred))
with(subset(Discontdataflex, treatment==1),lines(readingpre, pred,
  type = "l", lwd = "3"))
```

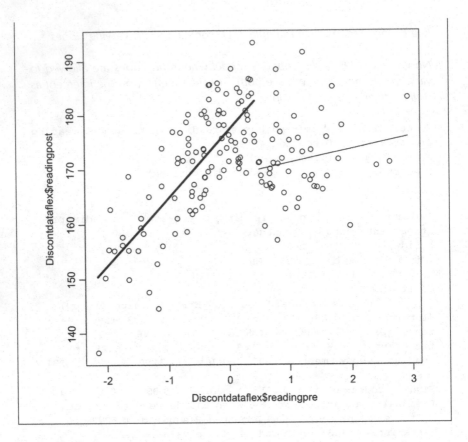

EXTENSION 3: MEDIATION

Motivating example: using the Perfectionism dataset. You are interested in the relationship between concern over mistakes and life satisfaction. You want to test whether the relationship between concern over mistakes and life satisfaction is explained by the self-esteem of the individual.

The key aspect of this research question is the desire to provide evidence for a proposed explanatory mechanism. In this case, self-esteem. Similar to moderating relationships, mediating relationships also describe the relationship between two variables and their relation to a third variable. In moderating relationships, the third variable changes the relationship between X and Y. In mediating relationships, the third variable instead explains the relationship between X and Y. This means that although the terms mediation and moderation are extremely similar and often confused, they imply very different conceptual relations and are tested in methodologically different ways.

To understand mediation, it is important to understand the different types of relationships between variables X (the independent variable), Y

(the outcome variable), and M (the moderator). We can first talk about the *total effect* of X on Y. The total effect is the effect of the independent variable on the outcome variable without taking the mediator into consideration. This means that if there is a mediating relationship, it has not yet been partialled out, and thus the effect of the mediator is still tied up in this effect. The total effect can be broken down into the direct effect of X on Y and the indirect effect of X on Y through M. The *direct effect* of X on Y is the effect of the independent variable on the outcome with the effect of the mediator removed. The *indirect effect* of X on Y through M, then, is the effect of the mediator on the outcome. Understanding these effects, we can see that a fully mediating relationship would be supported if there is a significant total effect, no significant direct effect, but a significant indirect effect. This would indicate that there is a relationship between X and Y, but this effect is completely explained by the mediator.

Historically, the recommendation was to test mediating relationships via a set of systematic steps involving correlation and hierarchical regression first put forth by Baron and Kenny (1986). These steps, however, only allow for the importance of the mediator to be inferred. Thus, more recent methodology provides significance tests for the indirect effect (mediator) which provides a stronger test for mediation. The following paragraphs will walk through first the original Baron and Kenny steps to test mediation and will then provide additional tests of the indirect effect.

Baron and Kenny (1986) Requirements for Testing Mediation

Baron and Kenny's 1986 work outlines four basic requirements for a mediating relationship. These four requirements work to support relationships between all relevant variables and then to demonstrate that the mediator explains the relationship between the independent variable and the outcome variable.

Requirement 1: there must be a relationship between the independent variable (X) and the dependent variable (Y).

Requirement 2: there must be a relationship between the independent variable (X) and the mediator (M).

Requirement 3: there must be a relationship between the mediator (M) and the outcome variable (Y).

Requirement 4: when both the independent variable and the mediator are used to predict the outcome, the relationship between the independent variable and the outcome becomes non-significant or is greatly reduced.

EXAMPLE: MEDIATION

Is the relationship between concern over mistakes and life satisfaction mediated by the self-esteem of the individual? Using Perfectionism data.

Requirements 1–3 can easily be supported through simple Pearson correlations.

```
library(Hmisc)
Meddata <-as.matrix(Perfectionism[c(8,59,9)])

rcorr(Meddata, type="pearson")
     LIF_SAT PEF_CM SE
LIF_SAT  1.00 -0.31 0.74
PEF_CM  -0.31  1.00 -0.40
SE       0.74 -0.40 1.00
n= 367
P
     LIF_SAT PEF_CM SE
LIF_SAT       0    0
PEF_CM   0         0
SE       0    0
```

Requirement 1 is met: the outcome variable (life satisfaction) is significantly correlated with the independent variable (concern over mistakes) (r=−.31, p<.05).

Requirement 2 is met: the outcome variable is significantly correlated with the mediator (self-esteem) (r=.74, p<.05).

Requirement 3 is met: the independent variable is significantly correlated with the mediator (r=−.4, p<.05).

Requirement 4 can be tested through the use of hierarchical regression adding the independent variable in one block and the mediator in a subsequent block. If, when the mediator is added, the independent variable becomes non-significant (or the beta reduces considerably) and the mediator is significant, a mediating relationship can be inferred.

```
med_model0 <- lm(LIF_SAT~PEF_CM, Perfectionism)
med_model1 <- lm(LIF_SAT~SE+PEF_CM, Perfectionism)
anova(med_model0, med_model1)
> summary(med_model0)
```

```
Call:
lm(formula = LIF _ SAT ~ PEF _ CM, data = Perfectionism)
Residuals:
 Min   1Q Median   3Q   Max
-3.4306 -0.5900 0.1055 0.6684 2.2246
Coefficients:
      Estimate Std. Error t value Pr(>|t|)
(Intercept) 6.05470  0.13813  43.83 < 2e-16 ***
PEF _ CM   -0.23404  0.03769  -6.21 1.45e-09 ***
---
Signif. codes: 0 '***' 0.001 '**' 0.01 '*' 0.05 '.' 0.1 ' ' 1
Residual standard error: 0.9141 on 365 degrees of freedom
Multiple R-squared: 0.09555, Adjusted R-squared: 0.09307
F-statistic: 38.56 on 1 and 365 DF, p-value: 1.446e-09

> summary(med _ model1)
Call:
lm(formula = LIF _ SAT ~ SE + PEF _ CM, data = Perfectionism)
Residuals:
 Min   1Q  Median   3Q   Max
-2.21641 -0.38486 0.05738 0.46003 1.65501
Coefficients:
      Estimate Std. Error t value Pr(>|t|)
(Intercept) 1.11966  0.27853  4.020 7.08e-05 ***
SE      0.76200  0.04025 18.934 < 2e-16 ***
PEF _ CM   -0.01029  0.02928 -0.351  0.725
---
Signif. codes: 0 '***' 0.001 '**' 0.01 '*' 0.05 '.' 0.1 ' ' 1
Residual standard error: 0.6497 on 364 degrees of freedom
Multiple R-squared: 0.5443, Adjusted R-squared: 0.5418
F-statistic: 217.4 on 2 and 364 DF, p-value: < 2.2e-16
>
> anova(med _ model0, med _ model1)
Analysis of Variance Table
Model 1: LIF _ SAT ~ PEF _ CM
Model 2: LIF _ SAT ~ SE + PEF _ CM
Res.Df RSS Df Sum of Sq  F  Pr(>F)
1  365 304.99
2  364 153.66 1  151.33 358.48 < 2.2e-16 ***
---
Signif. codes: 0 '***' 0.001 '**' 0.01 '*' 0.05 '.' 0.1 ' ' 1
```

> *When just the independent variable (concern over mistakes) is used to predict life satisfaction, concern over mistakes is a significant predictor (b = −.23, p<.001). When self-esteem is added to the model, now self-esteem is a significant predictor (b=.76, p<.001) and concern over mistakes is no longer significant (b=−.01, p=.725). Since the independent variable became non-significant once the mediator was added to the model, requirement 4 is met. With all four requirements met, we can say that self-esteem is supported as a mediator.*

Tests of Significance for the Indirect Effect

As mentioned above, the Baron and Kenny requirements provide a framework which will infer the presence of a mediating relationship through a pattern of significance. However, this framework does not provide a test of significance for the mediator. Since Baron and Kenny's publication, several methods have been suggested which provide a significance test for the mediator. Two popular methods are listed below:

The Sobel test. The Sobel test is a significance test which tests to see if the effect of the independent variable reduces significantly once the mediator is added into the regression model.

Bootstrapped confidence intervals. Recently, some authors have criticized the Sobel test for requiring a normal distribution for the sampling distribution of the indirect effect. For this reason, the use of bootstrapped confidence intervals for the indirect effect as an alternative. These confidence intervals allow for the same interpretation of significance as the Sobel test but do not run the risk of power loss due to non-normality.

MEDIATION EXAMPLE CONTINUED

Running the Sobel Test and Bootstrapped Confidence Intervals
The Sobel test can be run using the `mediation.test()` function from the {bda} package.

```
library(bda)
sobeltest         <-              mediation.test(Perfectionism$SE,
Perfectionism$PEF_CM, Perfectionism$LIF_SAT)

sobeltest
        Sobel     Aroian     Goodman
z.value -7.699559e+00 -7.690612e+00 -7.708538e+00
p.value 1.365363e-14 1.464331e-14 1.272671e-14 ·
```

This function gives the results of multiple tests in addition to Sobel. However, looking at the p value for Sobel we can see that the test is significant (p<.001) indicating that the effect of the independent variable reduced a significant amount when the mediator was introduced into the model. The indirect effect of the mediator is significant.

To obtain bootstrapped confidence intervals for the indirect effect, we can use the `mediate()` function from the {psych} package.

```
library(psych)
BootCI <- mediate(y=LIF _ SAT~PEF _ CM+(SE), data=Perfectionism)

Mediation/Moderation Analysis
Call: mediate(y = LIF _ SAT ~ PEF _ CM + (SE), data = Perfectionism)
The DV (Y) was LIF _ SAT. The IV (X) was PEF _ CM. The mediat-
ing variable(s) = SE.
Total   effect(c)   of PEF _ CM on LIF _ SAT = -0.23    S.E. = 0.04
t = -6.21 df= 365   with p = 1.4e-09
Direct effect  (c') of PEF _ CM on LIF _ SAT removing SE = -0.01
S.E. = 0.03 t = -0.35 df= 364   with p = 0.73
Indirect   effect   (ab)   of PEF _ CM on LIF _ SAT through SE
= -0.22
Mean bootstrapped indirect effect = -0.22 with standard error
= 0.03 Lower CI = -0.28  Upper CI = -0.17
R = 0.74 R2 = 0.54 F = 217.4 on 2 and 364 DF   p-value: 8.65e-81
```

This function will actually provide more feedback than just the bootstrapped confidence intervals. The output will break out the total effect and direct effect in addition to providing the bootstrapped confidence interval for the indirect effect. It will also provide the diagram below to help visualize the results. Notice that the slopes for the total effect and direct effect (also shown on the diagram) can be found on the original hierarchical regressions run to test Baron and Kenny requirement 4. The bootstrapped confidence interval for the indirect effect is [–.28, –.17]. Since zero is not in the confidence interval we can conclude that the indirect effect is significant. Putting everything together, with a significant total effect (p<.001), non-significant direct effect (p=.73), and significant indirect effect, we can say that mediation is supported in this analysis.

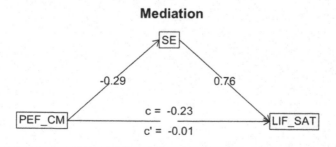

END OF CHAPTER SUMMARY

This chapter has provided three simple extensions of the multiple regression framework to address different types of research questions. Multiple regression is an extremely flexible framework that can test different hypotheses about different types of relationships and different types of research design situation and provide different types of information about an existing regression model. Clearly this chapter only brushed the surface with each of these topics, and the interested reader should see the listed resources for more detailed information. The next chapter of this book will present two more extensions of the multiple regression framework: modeling non-linearity and cross-validation procedures. It is fascinating to see just how far this single analysis can stretch in terms of utility and creativity!

RECOMMENDED RESOURCES

Cattaneo, M., Idrobo, N., & Titiunik, R. (2019). *A practical introduction to regression discontinuity designs: Foundations.* Cambridge: Cambridge University Press.
Hayes, A. (2018). *Introduction to mediation, moderation and conditional process analysis.* New York: Guilford Press.

CHAPTER 8: END OF CHAPTER EXERCISES

Using dataset chapter8ex1

1. Run a regression model testing the research questions: 'Does a student's gender moderate the relationship between age and attention? Does age moderate this relationship?'

Using dataset chapter8ex2

2. Run a regression discontinuity model to see if individuals receiving food assistance based on their low income have significantly lower food needs.

Using dataset chapter8ex3

3. Run a regression model testing whether self-control moderates the relationship between age and behavior.

Regression Extensions 2

Non-Linearity and Cross-Validation

THE PREVIOUS CHAPTER PRESENTED three extensions of the multiple regression framework which used dummy variables, interactions, and hierarchical regression to test research questions dealing with mediation, moderation, and regression discontinuity. The present chapter will present two more extensions of the multiple regression model which will allow for non-linear relationships to be modeled and present a framework for the cross-validation of predictive models. As will be detailed in the chapter, these topics make great sense to discuss together as the modeling of non-linear relationships can often lead to issues with model validity, and thus it is often important to cross-validate regression models with non-linear relationships.

EXTENSION 4: NON-LINEARITY

Chapter 4 discussed in depth the assumptions of multiple regression. One main assumption that has been taken for granted until this point has been the assumption of linearity. The assumption of linearity states that the relationship between predictors and the outcome needs to be linear. The reason behind this assumption is that the best fit line estimated by the multiple regression will be linear. If the relationships between the predictors and the outcome are not linear, the line of best fit will not well represent these relationships. When discussed in terms of correlation, other types of correlation (non-parametric correlations) were suggested as alternative

DOI: 10.1201/9780429295843-9

to the traditional Pearson correlation. For multiple regression, however, there are two potential avenues the researcher can take to still use OLS regression in the presence of non-linear relationships. These two options involve the use of (1) *variable transformations* or (2) *non-linear terms.*

Variable Transformations for Non-Linearity

When non-linear relationships exist in your data, one option is to *linearize the relationship* (flatten to a straight line) in order to meet the multiple regression assumption of linearity and to interpret all regression coefficients as would typically be done. Linearizing relationships is accomplished through the use of mathematical transformations applied to the variables which will maintain the interpretation of the variables and keep all cases in the same relative position but will cosmetically reshape the relationship to the desired shape. In this case, the desired shape would be a linear relationship. The most common transformations for linearizing relationships are the logarithmic function ($\log(x)$) and the square root function ($\sqrt{x}$).

Transformation Selection

When using transformations to flatten non-linearity, many options will exist. Will you transform both the outcome variable and the predictor? Will you transform just the outcome? Just the predictor? Will you use the logarithmic function or the square root function? There are several things the researcher may want to consider as they make their selection:

1. Transforming both the outcome and the predictor will generally produce the most linear solution.

2. Log transformations are often stronger than square root transformations.

3. Transformations are also not guaranteed to work or work well. Make sure you look at scatterplot after transforming to make sure that the selected transformation(s) were effective.

4. Transformed variables are still completely interpretable but the interpretation is sometimes more difficult than raw variables. Consider if you want to include multiple transformed variables.

5. If multiple non-linear relationships are present in a model, transforming the outcome variable is recommended because it will help alleviate non-linearity between each pair of variables.

Ultimately, transformation(s) chosen for your model should be effective in linearizing the relationship (as evidenced by the relationship visualized in the scatterplot) and should not hinder the interpretation of the model parameters. The researcher should feel free to try out different transformations to see which best linearizes the relationship. When doing so, however, make sure that the focus is on which best linearizes the relationship (visualized in the scatterplot) and not which most increases the variance explained by the model. Make sure model choices are based on assumptions and not fishing for good results.

What to Do with Negative Values?

Those familiar with mathematical functions may be wondering what to do if variables contain negative values. The natural logarithm or the square root of a negative value is undefined! Often variables will contain some negative values, so this can result in lost data (undefined values must be treated as missing) if not treated properly. In order to make sure no data are lost, a constant can be added to the variable prior to transforming in order to keep all values positive before transformation is applied. This will not impact the interpretation of the variable and will ensure that all values will be defined once the transformation is applied.

Pros and Cons to the Transformation Approach

As we will discuss two approaches to non-linearity, it is important to consider the benefits of each approach. The main benefit to using transformations to deal with non-linearity is that all regression parameters will remain interpretable. Use of transformations will not alter the meaning of the predictors used and will not introduce any unintended model complications (for example multicollinearity issues). Thus, this approach may be preferred by researchers creating regression models for conceptual explanation.

There are few major drawbacks to the use of transformations for linearizing relationships. It is important to realize, however, that transformations will have varying degrees of success. Depending on the shape and severity of linearity, there may not be a very successful transformation option. It is also important to remember that although transformations do leave the predictor able to be interpreted, the interpretation has changed slightly and the researcher needs to keep this in consideration as they interpret results.

EXAMPLE: TRANSFORMING FOR NON-LINEARITY

Motivating example: prediction of daily functioning as a function of age and gender. Lower values of daily functioning are an indication of more difficulty functioning. Gender has been dummy coded into one variable for male where 1=male and 0=female. The relationship between age and daily functioning is non-linear as evidenced by the scatterplot presented below. As a researcher you hope to conceptually learn about predictors of daily functioning. For this reason, you have chosen to linearize the relationship between daily functioning and age so that all model parameters can be interpreted.

```
plot(nonlinear$age, nonlinear$dailyfunction)
```

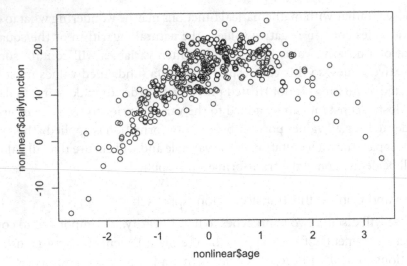

In order to linearize this relationship, we can consider different combinations of natural log and square root transformations for the outcome (daily functioning) and the predictor (age). Notice that values for both age and daily functioning have negative values (age has been standardized). For this reason, a constant (for age, 4, for daily functioning, 19) was selected which, when added, will make all values positive prior to transformation.

```
lnage<- log(nonlinear$age+4)
sqrtage<- sqrt(nonlinear$age+4)
lnfunction<- log(nonlinear$dailyfunction+19)
sqrtfunction<- sqrt(nonlinear$dailyfunction+19)
```

After looking at different combinations of transformations, the combination using the raw outcome (daily functioning) and the natural log of the predictor (age), appears the most successful combination for linearizing the relationship.

```
plot(lnage, nonlinear$dailyfunction)
```

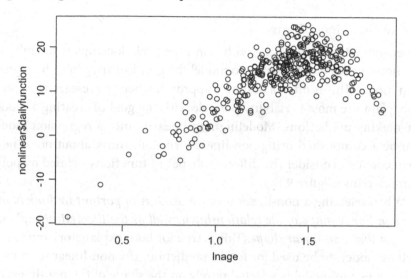

Having selected transformations to use, we can now run the regression model predicting daily function from log transformed age, and male.

```
transformed1<- lm(dailyfunction~lnage+male, nonlinear)
summary(lm.beta(transformed1))


Call:
lm(formula = dailyfunction ~ lnage + male, data = nonlinear)
Residuals:
Min     1Q  Median     3Q     Max
-16.4256 -2.2628  0.5053  2.8248  9.8815
Coefficients:
 Estimate Standardized Std. Error t value Pr(>|t|)
(Intercept) -9.1927     0.0000    1.2732  -7.220 3.26e-12 ***
lnage       16.4786     0.6885    0.9112  18.085 <2e-16 ***
male         2.2693     0.1609    0.5369   4.227 3.03e-05 ***
---
Signif. codes: 0 '***' 0.001 '**' 0.01 '*' 0.05 '.' 0.1 ' ' 1
Residual standard error: 4.471 on 350 degrees of freedom
Multiple R-squared: 0.4932, Adjusted R-squared: 0.4903
F-statistic: 170.3 on 2 and 350 DF, p-value: <2.2e-16
```

The regression model is significant (F = 170.3, p < .01) accounting for 49% of the variance in daily functioning. Both the log of age and male were *significant predictors of daily functioning with log of age being the strongest predictor. Both predictors are positive indicating that, as log of age increases, individuals have higher daily functioning and males have higher daily functioning than females.*

Use of Non-Linear Terms

A second option for dealing with non-linear relationships is to embrace the non-linearity and attempt to model the non-linearity into the regression model. This is often a desirable approach when the researcher wishes to explain the most variability possible with the goal of creating a model for making predictions. Modeling non-linearity into a regression model can be accomplished using non-linear terms. To think about non-linear terms, we can consider the different shapes of functions defined by polynomial terms (Figure 9.1).

When selecting a non-linear term to use *it is important to think about the non-linear shape of the relationship as well as the theoretical explanation for that non-linear shape.* This is true for both explanatory models as well as models to be used for future prediction. If a non-linear term used in a regression model is selected purely on the shape of the non-linearity with no connection to theoretical explanation of the non-linear shape, then the model estimated may be overfit. *Overfitting* refers to modeling too closely to the shape of the relationship in the sample resulting in poor generalizability to the population. Thus, whether you are attempting to explain a construct or attempting to predict a construct, overfitting will produce both problematic interpretations as well as problematic predictions. Once a non-linear term is selected for use in a regression model, it can simply be added mathematically as a slope in a similar fashion to interaction terms.

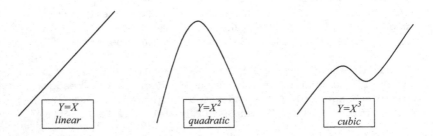

FIGURE 9.1   Polynomial functions.

Watch out for Multicollinearity!

Often, a non-linear term will be added to a model, rather than replacing the linear version of the term. This is because relationships are very rarely purely quadratic, or purely cubic, but rather a composite of functions. The creation of composite functions can be very beneficial when explaining variance; however, recall from the discussion on interaction terms that variables are very strongly related to variables they are part of. If you include time and time² both as predictors in your model, you will end up with a multicollinearity problem due to the strong correlation between these predictors. Make sure you are aware of this possibility as you are thinking through modeling options.

EXAMPLE: NON-LINEAR TERMS

Let us return to the prediction of daily functioning. As noted previously, the relationship between age and daily functioning is non-linear. In the previous example we chose to ignore this non-linearity and simply linearize it to remove it as an issue. However, conceptually, this non-linearity makes sense. For very young children, there will be much more difficulty with daily functioning (can't sit up/walk, can't use toilet, can't tie shoes, etc.). As children grow up they become much more self-sufficient indicating increasingly higher levels of daily functioning. As adults this level of daily functioning will continue to a point but likely begin to decrease as the individual enters old age. This conceptual description of the non-linear relationship would suggest a quadratic term for age should be added to the model.

```
plot(nonlinear$age, nonlinear$dailyfunction)
```

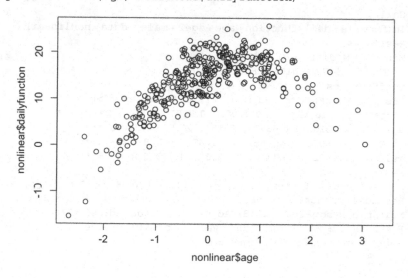

As a starting point, if the non-linearity is not modeled at all, the following model would result.

```
untransformed<- lm(dailyfunction~age+male, nonlinear)
summary(lm.beta(untransformed))
```

```
Call:
lm(formula = dailyfunction ~ age + male, data = nonlinear)
Residuals:
Min   1Q Median   3Q   Max
-34.653 -1.937  0.643  3.174  9.477
Coefficients:
  Estimate Standardized Std. Error t value Pr(>|t|)
(Intercept) 13.2633    0.0000    0.3364 39.430 <2e-16 ***
age       4.1121    0.6167    0.2831 14.527 <2e-16 ***
male      1.6203    0.1124    0.6118  2.648 0.00845 **
---
Signif. codes: 0 '***' 0.001 '**' 0.01 '*' 0.05 '.' 0.1 ' ' 1
Residual standard error: 5.249 on 350 degrees of freedom
Multiple R-squared: 0.3773,Adjusted R-squared: 0.3738
F-statistic:  106 on 2 and 350 DF, p-value:<2.2e-16
```

Once we add the quadratic term for age as the conceptual interpretation would suggest, the following model will result. Recall that to create a quadratic term, we simply take the product of the variable times itself (age*age).

```
nonlinear$age2 <- nonlinear$age*nonlinear$age
sqmodel <- lm(dailyfunction~age+age2+male, nonlinear)
summary(lm.beta(sqmodel))
```

```
Call:
lm(formula = dailyfunction ~ age + age2 + male, data = nonlinear)
Residuals:
Min   1Q Median   3Q   Max
-7.3136 -1.9472 0.0168 1.8565 8.0038
Coefficients:
  Estimate Standardized Std. Error t value Pr(>|t|)
(Intercept) 16.1282    0.0000    0.2087 77.263 <2e-16 ***
age       4.2510    0.6375    0.1545 27.523 <2e-16 ***
age2     -2.8972   -0.6621    0.1007 -28.768 <2e-16 ***
male      1.5992    0.1110    0.3337  4.793 2.44e-06 ***
---
Signif. codes: 0 '***' 0.001 '**' 0.01 '*' 0.05 '.' 0.1 ' ' 1
Residual standard error: 2.863 on 349 degrees of freedom
Multiple R-squared: 0.8153,Adjusted R-squared: 0.8137
F-statistic: 513.5 on 3 and 349 DF, p-value:<2.2e-16
>anova(untransformed, sqmodel)
```

```
Analysis of Variance Table
Model 1: dailyfunction ~ age + male
Model 2: dailyfunction ~ age + age2 + male
Res.Df  RSS  Df  Sum of Sq   F   Pr(>F)
1  350  9644.3
2  349  2860.6  1   6783.7  827.63 <2.2e-16 ***
---
Signif. codes: 0 '***' 0.001 '**' 0.01 '*' 0.05 '.' 0.1 ' ' 1
>
```

The model with no non-linear term accounted for 37% of the variance in daily functioning. Once we add the non-linear term, the variance explained jumps to 81%. Model comparison indicates this is a significant increase in variance explained. Like the previous model, we can say that age and male are significant positive predictors of daily functioning. However, the negative coefficient for the non-linear term has no conceptual explanation.

Pros and Cons to the Use of Non-Linear Terms

The main benefit to using non-linear terms is the ability to embrace the non-linear relationship observed in the data. This is of particular importance if the non-linear relationship has a clear conceptual explanation. The ability to model the non-linearity provides the opportunity to more closely model the relationships between variables and thus potentially account for higher portions of variance. This ability to more closely model the observed relationship can also be a drawback if used improperly. As mentioned previously, if the researcher uses non-linear terms to model too closely to the observed relationship, the model can become overfit to the sample and thus become a poor representation of the population.

A second drawback to the use of non-linear terms is coefficient interpretation. Non-linear terms have mathematical importance but can rarely be conceptually interpreted. A significant non-linear term can tell us that the non-linear term was important but doesn't have a conceptual explanation. It is also important to remember that inclusion of a non-linear term in addition to a linear term will often cause severe multicollinearity problems (due to the variable being nearly perfectly correlated with itself) and thus render the majority of the model uninterpretable. In general, then, the main utility of non-linear terms is found in the ability to produce models that account for high variance explained with less emphasis on conceptual interpretation.

EXTENSION 5: CROSS-VALIDATION

Motivating example: as a researcher you are working with a local school district to help administrators predict which students will be at risk for poor academic achievement. Your goal is to create a regression model that can be used to provide predictions of future math achievement. These predictions will be used as information for teachers to help anticipate which students will be in need of services.

The key element to this research situation is the practical usage of the generated regression equation. This situation wants to create a regression equation that can actually be used to help predict at risk kids. Thus, there are several elements that will be important.

1. *We want this model to account for a high amount of variance.* If the model predicts only weakly, it will be essentially useless in identification of the correct students.

2. *We want this model to predict well on samples it was not created on.* When regression parameters are estimated, they are estimated to maximize variance explained (minimize prediction error) for the specific dataset they are estimated from. This means that the parameters estimated will be optimal for that sample only. However, if a regression model is thoughtfully created based on theory, the estimated regression model should still work reasonably well for a sample from the same population.

3. *We want this model to be stable.* Based on the point above, when moving to a different sample from the same population, even when the model defined is exactly the same, the regression parameters will be different. However, if the model is created on a large enough sample (to maximize similarity to the population) and if the model is thoughtfully created based on theory, the deviation in the parameters from sample to sample should be minimal.

So, assuming that we have created a strong theory driven model of our construct, how do we validate that we have a *stable* theory driven model? This is the main driving motivation behind cross-validation. There are many approaches to cross-validation, but this text will briefly discuss two quick ways to assess the stability of a regression equation. The first method we will discuss (often called the validation approach) will create the regression equation and test the model fit on a new sample. The second method will run the same regression equation on two samples and compare parameters across the regression equations.

Cross-Validation Samples

In order to perform cross-validation on a regression model, the researcher must have two separate samples from the same population of interest. In some cases the researcher does have two separately collected samples from the same population. Often, however, this is achieved through acquiring one large sample from the population of interest and dividing it into two separate non-overlapping samples. The first of these samples will be the sample that is used to create the regression equation. This sample will be referred to as the *training sample*. Because the training sample is the sample that will create the actual regression equation for use, it is ideal for this sample to be as large as possible. Thus, if two different samples are not available and a single large sample must be cut in two, it is ideal that the training sample receive a larger proportion of cases than the *cross-validation sample*. The cross-validation sample is the sample that will be used to test the stability of the regression equation created in the training sample.

Cross-Validation Procedures

When performing a cross-validation, we start with creating the regression equation on the training sample. In practice this step may have already occurred in a previously conducted study where the focus of the study was creating a strong prediction model for the construct of interest. Once the regression equation is created on the training sample, the focus becomes assessing the stability of the equation. The most common method for assessing model stability is using the prediction model to predict the construct of interest in a new sample and computing the cross-validation R squared. The cross-validation R squared is a measure of how much variance the regression equation accounts for in a sample other than the one it was created on. This cross-validation R squared can then be compared to the R squared in the training sample for a measure of *shrinkage*. Some degree of shrinkage is always to be expected since the parameters of the regression equation are only optimized for the training sample. However only a small amount of shrinkage should occur if the model was created based on sound theory and knowledge of the population. If the researcher models too closely to the visual trends and nuances of the sample instead of thinking in terms of the entire population the model can become *overfit* to the training sample and fit shrinks considerably when used on the cross-validation sample.

A second quick method the researcher may choose to use in addition to computation of the cross-validation R squared is to quickly run the same regression model on the cross-validation sample. The purpose is not to use the second generated regression equation, but rather to compare parameter values and significance to see if similar patterns emerge. For a stable model we would expect to see all parameters in the same direction (if positive in the training sample, should also be positive in the cross-validation sample), similar order of relative strength, and the same predictors should still be significant.

EXAMPLE: R CROSS-VALIDATION R SQUARED

Using the motivating example above: we will create and cross-validate a regression model predicting student achievement from their midterm grades on reading, math, writing, and the student's gender. For this example a large dataset (Achieve) was cut into two non-overlapping samples. achieve_sample1 (N = 500) will be used as the training sample and achieve_sample2 (N = 500) will be used to cross-validate.

Step 1: create the regression equation on the training sample.

```
trainingmodel <-lm(achieve~r _ midterm+m _ midterm+w _ mid-
term+male,  data=achieve _ sample1)
summary(trainingmodel)
```

```
Call:
lm(formula=achieve~r _ midterm+m _ midterm+w _ midterm+male,
data=achieve _ sample1)
Residuals:
Min   1Q Median   3Q   Max
-340.59 -68.94  -0.89  63.50 362.94
Coefficients:
 Estimate Std. Error t value Pr(>|t|)
(Intercept) 134.69100  36.47520  3.693 0.000247 ***
r _ midterm  -11.84215  0.51552 -22.971 <2e-16 ***
m _ midterm   5.95773  0.09217 64.640 <2e-16 ***
w _ midterm   3.08487  0.13669 22.569 <2e-16 ***
male     -8.00922  9.83569 -0.814 0.415863
---
Signif. codes: 0 '***' 0.001 '**' 0.01 '*' 0.05 '.' 0.1 ' ' 1
Residual standard error: 104.8 on 495 degrees of freedom
Multiple R-squared: 0.9118, Adjusted R-squared: 0.9111
F-statistic: 1280 on 4 and 495 DF, p-value: <2.2e-16
```

Note, this model accounts for approximately 91% of the variance in student achievement. Given this is a model we actually want to use to make predictions it is very important the variance explained be very high.

Step 2: use the training equation to make predictions in the cross-validation sample and compute the cross-validation R squared.
    One very simple way to achieve this is to directly use the regression equation and plug in the values from the cross-validation sample. Below you can see the results for the first seven cases.

```
predictions<-  134.691-11.84215*(achieve _ sample2$r _ midterm)+
5.95773*(achieve _ sample2$m _ midterm)+
3.08487*(achieve _ sample2$w _ midterm)-8.00922*(achieve _ sa
mple2$male)
```

```
501     502     503     504     505     506    507
1151.00973                              1136.62937 681.57257 893.54024
1193.34103 570.22818 395.68281
```

An easier way to achieve this is to use the predict() function. The same first seven cases are printed below. They differ by decimal places due to rounding but it is clear that these functions achieve the same purpose.

```
predictions _ easy<-   trainingmodel  %>%  predict(achieve _
sample2)
```

```
[1]  1151.00962  1136.62933 681.57253 893.54030  1193.34100
570.22826 395.68298
```

Once the predictions are obtained they can be used to compute the cross-validation R squared.

```
R _ squared=R2(predictions _ easy,achieve _ sample2$achieve)
R _ squared
[1]  0.9072765
```

The equation accounts for approximately 90% of the variance in student achievement when used on the cross-validation sample. As expected, this is less than the 91% of the variance accounted for in the training sample where the equation was created. However a shrinkage of 1% is very little, indicating this equation is likely stable across samples.

EXAMPLE: COEFFICIENT STABILITY

Continuing with the previous cross-validation example, we will now run the regression equation a second time, this time on the cross-validation sample, in order to compare stability of the regression coefficients.

Recall the output obtained from the training model. This time the lm.beta() function was used so the strength of predictors can be compared using the standardized coefficients.

```
trainingmodel <-lm(achieve~r _ midterm+m _ midterm+w _ mid-
term+male,  data=achieve _ sample1)
summary(trainingmodel)


Call:
lm(formula = achieve ~ r _ midterm+m _ midterm+w _ midterm+male,
data = achieve _ sample1)
Residuals:
Min   1Q Median   3Q    Max
-340.59 -68.94  -0.89   63.50 362.94
Coefficients:
 Estimate Standardized Std. Error t value Pr(>|t|)
(Intercept) 134.69100   0.00000  36.47520  3.693 0.000247 ***
r _ midterm  -11.84215   -0.30659  0.51552 -22.971 <2e-16 ***
m _ midterm  5.95773   0.86459  0.09217 64.640 <2e-16 ***
w _ midterm  3.08487   0.30151  0.13669 22.569 <2e-16 ***
male    -8.00922   -0.01089  9.83569 -0.814 0.415863
---
Signif. codes: 0 '***' 0.001 '**' 0.01 '*' 0.05 '.' 0.1 ' ' 1
Residual standard error: 104.8 on 495 degrees of freedom
Multiple R-squared: 0.9118, Adjusted R-squared: 0.9111
F-statistic: 1280 on 4 and 495 DF, p-value: <2.2e-16
```

We will now run the same model again but using achieve_sample2.

```
crossmodel <-lm(achieve~r _ midterm+m _ midterm+w _ mid-
term+male,  data=achieve _ sample2)
summary(lm.beta(crossmodel))


Call:
lm(formula = achieve ~ r _ midterm+m _ midterm+w _ midterm+male,
data = achieve _ sample2)
Residuals:
Min   1Q Median   3Q    Max
-426.23 -70.81  -5.96   73.64 268.77
Coefficients:
 Estimate Standardized Std. Error t value Pr(>|t|)
(Intercept) 108.71237   0.00000  33.91441  3.205 0.00144 **
```

```
r_midterm  -10.71483  -0.30129  0.48549 -22.070 <2e-16 ***
m_midterm    5.88603   0.88614  0.09094  64.722 <2e-16 ***
w_midterm    2.94922   0.32195  0.12530  23.536 <2e-16 ***
male       -10.84635  -0.01487  9.95143  -1.090 0.27627
---
Signif. codes: 0 '***' 0.001 '**' 0.01 '*' 0.05 '.' 0.1 ' ' 1
Residual standard error: 102.5 on 495 degrees of freedom
Multiple R-squared: 0.908, Adjusted R-squared: 0.9073
F-statistic: 1222 on 4 and 495 DF, p-value: <2.2e-16
```

Comparing models:

1. *R squared of the two models is very similar (91% in training compared to 90% in cross).*
2. *All parameters are in the same direction in both models. R_midterm and male are negative in both models. M_midterm and w_midterm are positive in both models.*
3. *All parameters except male are significant in both models.*
4. *m_midterm is the strongest predictor in both models followed by r_midterm and w_midterm which are approximately equal and male is the weakest.*

Comparison of parameter direction, strength, and significance indicates that this model is stable across samples.

END OF CHAPTER SUMMARY

This chapter talked about two regression extensions: dealing with non linearity and cross-validating models. Although these two topics on their own lend much to the research design toolbox, the savvy reader will have realized, these two topics were placed together very deliberately. The modeling of non-linearity can lead to potential issues of overfitting. Cross-validation is a way to identify when overfitting has occurred. The final chapter will close the book out with two last extensions of the multiple regression model to situations of nested data. The nested data extensions will pull together all of the topics discussed in this text to provide a glimpse of what further extensions are possible.

CHAPTER 9: END OF CHAPTER EXERCISES

Using dataset chapter9_train

1. Run a scatterplot of the relationship between age and expense.

2. Use transformations to linearize the relationship between age and expense. Then run a regression model predicting expense from transformed age and male.

3. Run a regression model predicting expense from male, age, and a quadratic term for age.

Using dataset chapter9_cross

4. Run a cross-validation R squared on the regression model predicting expense from male, age, and the quadratic term for age. How much does the R squared shrink in the new sample?

5. Run the model predicting expense from male, age, and the quadratic term for age on the cross-validation sample. How stable are your model coefficients?

Regression Extensions 3

Nested Data

I N THIS FINAL CHAPTER, we will extend the multiple regression frame-work one more time, this time allowing for nested levels of influence to be accounted for. Nested data or nested levels of influence occur when your units of interest are nested or contained within higher level units, or context. For example: consider the test anxiety data we have studied several times over the course of this book. This sample was collected from one mid-sized public university. Since all units (undergraduate students) are from the same context, there are no nested levels to worry about. However, if instead, this dataset included undergraduate students from 20 different universities across the country, we now have to deal with accounting for which college the student attended in addition to just accounting for characteristics of the student. This would be an example of a two-level nested sample where undergraduates are nested within university.

In this chapter we will learn two different regression-based methodologies for dealing with nested samples. The first methodology, *fixed effects modeling*, is a simple option best at dealing with small numbers of level 2 (context) units and simpler modeling options. The second methodology, *hierarchical linear modeling*, is a much more complex option which provides modeling options for many types of nested designs and complex modeling options. It should be noted that with each of these methodologies there is far more that can be done than will be presented here. This chapter is meant as an introduction to each of these options. Resources

DOI: 10.1201/9780429295843-10

will be provided for those who wish to pursue each of these options more fully.

FIXED EFFECTS MODELING

Fixed effects modeling is a very simple method that uses dummy variables to account for context nesting structure. At its most basic, fixed effects modeling is just running a normal multiple regression model but incorporating dummy variables to represent the level 2 nesting units. Due to the simplicity of this method, there are a number of customizations presented in current literature, which can expand this method to allow for more complex data situations (for example, the inclusion of level 2 predictors); however, at its core, this method is best for simple nested data situations with a small number of level 2 units.

EXAMPLE: FIXED EFFECTS MODELING USING R USING PT_FEM DATA

This example uses a subset of a school achievement dataset. This dataset includes a sample of 592 students nested within 11 different schools. The purpose of the present analysis is to predict students' measure of total achievement from measures of memory, non-verbal skills, and age. However, since the students are nested within 11 different schools, dummy variables will be used to account for the variance due to school level nesting structure.

To begin, dummy variables need to be created for the 11 schools. Chapter 5 used if/else statements to create dummy variables. Since FEM generally requires a larger number of dummy variables to be created, we will instead use the dummy _ cols() from the {fastDummies} package. This function will create a new dataset with all dummy variables using one line of code and each dummy will have the same root based on the original coding of the column. It is recommended that the dataset be renamed so that the original does not get overwritten.

```
library(fastDummies)
library(lm.beta)
PT _ FEM <- dummy _ cols(PTedited, select _ columns="school")
```

Once dummy variables are created, the regression can be run using the same code described in the previous chapters of this text. Recall that only $k - 1$ dummy variables should be used in the model, as use of all dummied levels will produce complete collinearity. Thus, only dummy variables for schools 1–10 will be included.

```
FEM _ example <- lm(getotal~npamem+npaverb+age
    +school _ 1+school _ 2+school _ 3+school _ 4+school _ 5+scho
ol _ 6
```

```
      +school _ 7+school _ 8+school _ 9+school _ 10,PT _ FEM)
summary(lm.beta(FEM _ example))

Call:
lm(formula = getotal ~ npamem + npaverb + age + school _ 1 +
school _ 2 +
school _ 3 + school _ 4 + school _ 5 + school _ 6 + school _ 7 +
school _ 8 +
school _ 9 + school _ 10, data = PT _ FEM)
Residuals:
Min   1Q Median  3Q   Max
-2.4889 -0.8333 -0.1557 0.6694 3.7777
Coefficients:
    Estimate Standardized Std. Error t value Pr(>|t|)
(Intercept) -1.945354  0.000000  1.201473 -1.619 0.105985
npamem     0.005691  0.103922  0.001901  2.994 0.002880 **
npaverb    0.035260  0.603381  0.002097 16.814 < 2e-16 ***
age      0.035258  0.112931  0.010695  3.297 0.001041 **
school _ 1  0.323710  0.065196  0.225481  1.436 0.151667
school _ 2  0.246362  0.036067  0.276033  0.893 0.372507
school _ 3  0.020897  0.004365  0.221965  0.094 0.925027
school _ 4  0.268184  0.045107  0.251901  1.065 0.287501
school _ 5 -0.094235 -0.022681  0.207676 -0.454 0.650181
school _ 6  0.495291  0.109608  0.216867  2.284 0.022757 *
school _ 7  0.274046  0.053464  0.230886  1.187 0.235762
school _ 8  0.465286  0.082730  0.242274  1.920 0.055307.
school _ 9  0.996952  0.136013  0.288282  3.458 0.000585 ***
school _ 10 0.272353  0.049420  0.239007  1.140 0.254978
---
Signif. codes: 0 '***' 0.001 '**' 0.01 '*' 0.05 '.' 0.1 ' ' 1
Residual standard error: 1.174 on 555 degrees of freedom
(23 observations deleted due to missingness)
Multiple R-squared: 0.4047,Adjusted R-squared: 0.3908
F-statistic: 29.03 on 13 and 555 DF, p-value: < 2.2e-16
```

This model using memory, verbal, age, and school accounts for 40.47% of the variance in total achievement score. All three predictors of interest, memory (β = .1039), verbal (β = .603), and age (β = .1139), are significant positive predictors of total achievement. Verbal, with a beta of .601, was the strongest predictor of total achievement. The dummy variables included for school are there to account for school level variance but do not get interpreted. If we want to see how much variance knowledge of school attended accounts for we can run a second model with just the predictors of interest (memory, verbal, and age) and compare the models to see the difference in variance accounted for.

```
Call:
lm(formula = getotal ~ npamem + npaverb + age, data = PT _ FEM)
Residuals:
Min    1Q Median   3Q   Max
-2.8116 -0.8370 -0.2082 0.7188 3.9092
Coefficients:
     Estimate Standardized Std. Error t value Pr(>|t|)
(Intercept) -1.700944   0.000000  1.195946 -1.422 0.15550
npamem     0.005831   0.106489  0.001853  3.147 0.00174 **
npaverb    0.035261   0.603399  0.002046 17.236 < 2e-16 ***
age        0.035162   0.112622  0.010751  3.270 0.00114 **
---
Signif. codes: 0 '***' 0.001 '**' 0.01 '*' 0.05 '.' 0.1 ' ' 1
Residual standard error: 1.191 on 565 degrees of freedom
(23 observations deleted due to missingness)
Multiple R-squared: 0.3755, Adjusted R-squared: 0.3721
F-statistic: 113.2 on 3 and 565 DF, p-value: < 2.2e-16


> anova(FEM _ model2, FEM _ model1)
Analysis of Variance Table
Model 1: getotal ~ npamem + npaverb + age
Model 2: getotal ~ npamem + npaverb + age + school _ 1 + school _ 2
+ school _ 3 +
school _ 4 + school _ 5 + school _ 6 + school _ 7 + school _ 8 +
school _ 9 +
school _ 10
Res.Df  RSS Df Sum of Sq  F  Pr(>F)
1  565  802.00
2  555  764.41 10  37.581 2.7285 0.002774 **
---
Signif. codes: 0 '***' 0.001 '**' 0.01 '*' 0.05 '.' 0.1 ' ' 1
>
```

This second model only accounts for 37.55% of the variance in total achievement. This is a difference of 2.92%. Using the anova() command to compare the two models we find that adding the dummy variables to account for school level nested structure significantly increased the variance explained.

HIERARCHICAL LINEAR MODELING

Motivating example: your goal is to find predictors of math achievement for elementary school aged children. You want to consider characteristics of the student in your prediction of math achievement as well as characteristics of the school the student attends. The sample contains 10,898 students nested within 163 schools.

To reiterate: what makes this situation different from traditional multiple regression is that this sample is nested. Each child in this sample belongs to a particular elementary school. Or in other words, children are nested within schools. Children being nested within different contexts (schools) means that the context of each child needs to be taken into account. Ignoring context when analyzing data can lead to incorrect and even contradictory results. Thus, running a multiple regression will not be sufficient in this situation. In the previous section we talked about one potential option for nested data, fixed effects modeling, where dummy variables could be used to account for context level variance. However, in the previous example, we were not interested in any sort of context (school) level predictors. Because this example explicitly states school level characteristics are to be considered as well as student level predictors, we may want to work with a different analysis which can more flexibly handle predictors at different levels of analysis.

Hierarchical linear modeling is a multiple regression framework that accounts for context level information by using nested regression models. Mathematically, instead of just having one regression equation, there are different regression equations at each level of analysis. The basic two-level form for a hierarchical linear model can be defined as follows:

Level 1: $Y_{ij} = \beta_{oj} + \beta_{1j}\left(X_{ij}\right) + e_{ij}$ (10.1)

Where:
   Y_{ij} is the outcome variable for the ith individual in the jth group.
   X_{ij} is a level 1 predictor variable for the ith individual in the jth group.
   β_{oj} is the intercept of the level 1 equation. It is also equivalent to the grand mean on the outcome.
   β_{1j} is the slope of a level 1 predictor variable: the change in the outcome for every one unit change in the level 1 predictor.
   e_{ij} is the error in prediction for the ith individual in the jth group.

The level 1 equation is essentially the linear multiple regression model predicting a level 1 outcome from level 1 characteristics (for example predicting student math achievement from characteristics of the student). Moving from level 1 to level 2, the structure of the equations will be the same but the outcome will no longer be the outcome variable. Instead, at level 2, the outcomes to be predicted will be the level 1 parameters (intercept and slopes). The resulting framework essentially defines a separate multiple regression model predicting the outcome variable for each level 2 unit.

Level 2: $\beta_{0j} = \gamma_{00} + \gamma_{01}(W_j) + u_{oj}$

$$\beta_{1j} = \gamma_{10} + \gamma_{11}(W_j) + u_{1j} \tag{10.2}$$

Where:

γ_{00} is the intercept of the level 2 equation predicting the intercept. This is also mathematically equivalent to the grand mean of the entire multilevel model.

γ_{01} is the slope for a level 2 characteristic predicting the level 1 intercept. This will be mathematically equivalent to a level 2 characteristic predicting the outcome variable.

W_j is a level 2 predictor variable for the jth group.

u_{oj} is the random effect for the level 1 intercept or the error in predicting the level 1 intercept.

u_{1j} is the random effect for a level 1 slope or the error in predicting the level 1 slope.

The level 2 equations essentially represent the value of accounting for level 2 structure and level 2 characteristics to the level 1 model. Although at first glance this may look very different from the multiple regression framework we have been working with, if we use simple substitution to substitute the level 2 equations into the level 1 model we obtain the following:

$$Y_{ij} = \left[\gamma_{00} + \gamma_{01}(W) + u_{0j}\right] + \left[\gamma_{10} + \gamma_{11}(W) + u_{1j}\right](X_{ij}) + e_{ij} \tag{10.3}$$

Which can then be simplified and rearranged to obtain what is called the full model:

$$Y_{ij} = \gamma_{00} + \gamma_{01}(W_j) + \gamma_{10}(X_{ij}) + \gamma_{11}(X_{ij})(W_j) + u_{0j} + u_{1j}(X_{ij}) + e_{ij} \tag{10.4}$$

From the full model several things become apparent.

1. The parameters being estimated by the HLM model will be the gammas (γ_{00}, γ_{01}, γ_{10}, γ_{11}): these will be known as the *fixed effects*. And the error terms (u_{0j}, u_{1j}, and e_{ij}): these will be known as the *random effects*.

2. The fixed effects model is essentially the same as the multiple linear regression model described in the previous chapters of this text. γ_{00} is the intercept of the overall multilevel model. γ_{10} and γ_{01} are both slope coefficients for level 1 and level 2 predictors and will be interpreted in the same manner as slopes in multiple linear regression.

3. γ_{11} will represent the slope for a *cross-level interaction*. Cross-level interactions happen when the interaction between a lower-level predictor and a higher-level predictor is modeled in to the design. These interactions can be interpreted in exactly the same way as interactions in the multiple linear regression framework; however they have the added interpretation as representing the effect of context on the importance of lower-level predictors.

Random Effects and the Tau Matrix

The random part of the model is what makes the multilevel framework special. Although modeled as errors in the model, u_{0j} and u_{1j} are not actual error but rather represent the degree that the level 2 context units differ from one another. When the model is estimated, random effects (u_{0j}, u_{1j}...) will be estimated as a matrix called the *tau matrix*. The tau matrix contains the variances and covariances between the random effects.

$$\tau = \begin{bmatrix} \tau_{00} & \\ \tau_{10} & \tau_{11} \end{bmatrix} \tag{10.5}$$

Where

τ_{00} is the variance of u_{0j}

τ_{10} is the variance of u_{1j}

τ_{11} is the covariance of u_{0j} and u_{1j}

HLM estimates separate level 1 regression lines for each level 2 unit (for example, estimates a regression line predicting math ability from student characteristics for each different school in the sample). The parameters to be interpreted, however, are essentially averages across the parameters for each regression line. It is the random effects that provide us with information about how different the intercepts and slopes of these regression lines for the different units are from each other.

HLM Using R Software

Hierarchical linear models can be estimated in R using the lmer() function in the {lme4} package. The `lmer()` function requires definition of the fixed and random parts of the model, and the data that the model will use. Model definition is very similar to that of the `lm()` model used for regression.

EXAMPLE 1: STARTING SIMPLE. THE NULL MODEL USING ACHIEVE DATA

Not all HLM models have all the parameters defined in the previous section. The simplest HLM model is the model called the *null model*. The null model is the model that predicts the outcome from the intercept only. This model is used mainly to help compute effect size measures and model comparisons (see Bolin, Finch, and Kelley 2019 for further information on these topics). Using the motivating example at the beginning of the chapter, we will begin by creating a null model predicting reading achievement.

Level 1: $Math_{ij} = \beta_{0j} + e_{ij}$

Level 2: $\beta_{0j} = \gamma_{00} + u_{0j}$ (10.6)

Using the `lmer()` function we need to define the fixed and random parts of the model. The fixed part of the model will be geread~1. This can be read as *geread is predicted by the intercept* (in this code 1 stands for the intercept). Next, the random part of the model is added to the fixed part in parentheses (1|school). This can be read as the *intercept is random across schools*. This part of the code both designates the random effects present in the model (here only a random intercept) and also indicates the nesting structure of the model (individuals are nested within schools).

```
library(lme4)
null <- lmer(geread~1+(1|school), Achieve)
summary(null)
Linear mixed model fit by REML ['lmerMod']
Formula: geread ~ 1 + (1 | school)
Data: Achieve
REML criterion at convergence: 48951.5
Scaled residuals:
Min   1Q Median  3Q   Max
-2.3389 -0.6366 -0.2131 0.2860 3.8732
```

```
Random effects:
Groups   Name     Variance Std.Dev.
school   (Intercept) 0.4225  0.650
Residual          5.0753  2.253
Number of obs: 10903, groups: school, 163
Fixed effects:
    Estimate Std. Error t value
(Intercept)  4.319   0.056   77.12
```

Using the summary() command we can see the output for this model. At the top a summary of the residuals is provided for help with the identifying of outliers. Below is a summary of the variance of the random effects. The variance of the intercept $\tau_{00} = .4225$ and the overall error variance is 5.0753. Under the fixed effects, we can find the estimate of the intercept ($\gamma_{00} = 4.319$, $t = 77.12$). This t value is significant at $p < .05$ indicating that the intercept is significantly different from zero.

EXAMPLE 2: ADDING COMPLEXITY. RANDOM INTERCEPTS MODELS

Now will estimate a more interesting model. Using the motivating example at the beginning of the chapter, we will now create a *random intercepts model* (model with only a random intercept but no random slopes) predicting reading achievement from vocabulary (individual level characteristic) and school enrollment (level 2 school level characteristic).

Level 1: $Math_{ij} = \beta_{0j} + \beta_{1j}(gevocab) + e_{ij}$

Level 2: $\beta_{0j} = \gamma_{00} + \gamma_{01}(senroll) + u_{0j}$ (10.7)

$\beta_{1j} = \gamma_{10}$

```
library(lme4)
RandomIntercepts <- lmer(geread~gender+ senroll+
(1|school), Achieve)
summary(RandomIntercepts)
Linear mixed model fit by REML ['lmerMod']
Formula: geread ~ gevocab + senroll + (1 | school)
Data: Achieve
REML criterion at convergence: 45690.9
Scaled residuals:
Min   1Q Median   3Q   Max
-3.0574 -0.5715 -0.2110 0.3182 4.4141
```

```
Random effects:
Groups   Name      Variance Std.Dev.
school   (Intercept) 0.1081  0.3287
Residual          3.8078  1.9514
Number of obs: 10898, groups: school, 163
Fixed effects:
    Estimate Std. Error t value
(Intercept) 2.045e+00 1.150e-01 17.783
gevocab   5.089e-01 8.158e-03 62.380
senroll   9.884e-07 2.080e-04  0.005
Correlation of Fixed Effects:
 (Intr) gevocb
gevocab -0.312
senroll -0.905 -0.007
```

First, notice that the code for defining the model did not need to differentiate between level 1 and level 2 predictors. The fixed effects portion of the code looks almost the same as in the lm() function for linear regression. Printed first are the random effects, providing variance estimates for the random intercept (τ_{00} = .1081) and the overall residual error variance (3.8078). The fixed effects portion of the output reads very much like the intercept and slope coefficients of a linear regression model. The overall intercept is significant (γ_{00} = .2.045, t = 17.783) indicating that the intercept is significantly different from zero. The slope for gevocab is also significant (γ_{10} = .5089, t = 62.38) indicating that vocabulary is a significant, positive predictor of reading ability. The slope for school enrollment was not significant (γ_{01} < .000, t = .005) indicating that school enrollment does not significantly predict reading ability.

EXAMPLE 3: ADDING FURTHER COMPLEXITY. RANDOM COEFFICIENTS MODELS AND CROSS-LEVEL INTERACTIONS

As a final example, we will now add the cross-level interaction between gender and school enrollment to the model and allow the impact of gender to be random. Allowing any of the level 1 slopes to be random designates the model as a *random coefficients model*. In the R code for this model, the cross-level interaction is designated as the product of the two interacting variables (gevocab*senroll). To designate the slope for gender as random, the random part of the code can be changed to (gevocab|school) to indicate that the slope for gender will be allowed to vary randomly in addition to the intercept.

Level 1: $Math_{ij} = \beta_{0j} + \beta_{1j}(gevocab) + e_{ij}$

Level 2: $\beta_{0j} = \gamma_{00} + \gamma_{01}(senroll) + u_{0j}$ (10.8)

$\beta_{1j} = \gamma_{10} + \gamma_{11}(senroll) + u_{1j}$

```
library(lme4)
RandomCoef <- lmer(geread~gevocab+senroll+gevocab*senr
oll+ (gevocab|school), Achieve)
summary(RandomCoef)
Linear mixed model fit by REML ['lmerMod']
Formula: geread ~ gevocab + senroll + gevocab * senroll +
(gevocab | school)
Data: Achieve
REML criterion at convergence: 45566.7
Scaled residuals:
Min   1Q Median   3Q   Max
-3.6902 -0.5679 -0.2088 0.3136 4.6108
Random effects:
Groups  Name    Variance Std.Dev. Corr
school  (Intercept) 0.31916 0.5649
   gevocab   0.01888 0.1374  -0.84
Residual       3.70895 1.9259
Number of obs: 10898, groups: school, 163
Fixed effects:
     Estimate Std. Error t value
(Intercept)  1.895e+00 2.135e-01  8.874
gevocab   5.376e-01 4.772e-02 11.267
senroll   2.601e-04 4.022e-04  0.647
gevocab:senroll -4.014e-05 9.046e-05 -0.444
Correlation of Fixed Effects:
   (Intr) gevocb senrll
gevocab   -0.865
senroll   -0.957 0.829
gevcb:snrll 0.824 -0.955 -0.863
```

Beginning with the random effects, variance estimates are provided for the variance of the random intercept ($\tau_{00} = .319$), and the variance of the random slope for gevocab ($\tau_{10} = .0189$) as well as the overall residual error variance (3.709). Notice across examples 1, 2, and 3, the residual error variance reduced with the addition of predictors and model elements clearly demonstrating how the additional components are helping to explain the variance in geread. The random effects output also provides an estimate for the covariation between the random intercept and random slope ($\tau_{11} = -.84$) indicating a strong relationship between random intercept and the random slope. Looking to the fixed effects, the intercept and the slope for *vocabulary are still significant predictors of reading ability. The slopes for school enrollment and the cross-level interaction between school enrollment and vocabulary (γ_{11} < .001, t = −.444) are not significant indicating the school enrollment is not a significant predictor of reading ability and school enrollment does not change the impact of vocabulary on reading ability.*

CONCLUDING COMMENTS ON HIERARCHICAL LINEAR MODELING

The presentation of HLM in this text is meant to provide an introduction to the topic as well as demonstrate how HLM is really an extension of the multiple linear regression framework. As is likely evident at this point, HLM and multilevel modeling in general is a much larger topic than will be covered here. Once the basic HLM framework and R code are understood there is much to discuss in terms of assumptions, model building, model comparison, estimation options, and extensions to other data types. Essentially, the entire structure of this text could then be applied to hierarchical linear modeling! This text will not explore these topics with respect to hierarchical linear modeling; however readers are urged to see Finch, Bolin, and Kelly (2019) for an in depth look at these topics using R software.

SUMMARY

This final chapter concludes this multiple regression text with extensions of the multiple linear regression framework for nested data situations. Fixed effects modeling provides a simple modeling solution for simple two-level nested data situations whereas hierarchical linear modeling provides a more flexible modeling solution for more complex nested data situations. These extensions are an exciting way to end this text as the reader can easily conceptualize how each of the topics presented in the previous chapters could be applied to these nested data situations. In essence all of the content in this text can be pulled together to create well thought out models tailored well to the theoretical and practical considerations of the research situation. The multiple regression framework has many potential applications and it is exciting to think about how it will be used next!

RECOMMENDED RESOURCES

Allison, P. (2009). *Fixed effects regression models.* Sage Publications.

Finch, W., Bolin, J., & Kelley, K. (2019). *Multilevel modeling using R* (2nd ed.). Boca Raton, FL: CRC Press.

CHAPTER 10: END OF CHAPTER EXERCISES

Using dataset chapter10ex1

1. Run a fixed effects model predicting getotal from npamem, gereadcm, and age using dummy variables to account for school level variance. Does accounting for school level variance significantly increase variance explained?

Using dataset chapter10ex2

2. Run a hierarchical linear model null model predicting getotal from an intercept only with the nesting structure being students nested within classrooms.

3. Run a hierarchical linear random intercepts model predicting getotal from npamem, gereadcm, and ptratio.

4. Run a hierarchical linear random coefficients model predicting getotal from npamem, gereadcm, and ptratio with the slope for gereadcm allowed to vary across classrooms. Interpret your model. Which model coefficients are significant?

Appendix A: Introduction to R

As mentioned in the introduction, the choice to use R software for this textbook was predicated on the flexibility and power of the program as well as the added benefit of being a free download. Although R software is becoming much more ubiquitous among applied researchers, it is recognized that not all applied researchers have had good introductory experiences with R. This appendix provides an introduction to the R platform and basic operations in R to get the researcher who has never used R before proficient with basic R use.

R SHELLS

As mentioned above, basic R software consists of a single blank console window. The user can type directly into the console window or create script files which are then run in the console window. There have been many free shell programs created for R which aim to improve upon and simplify the experience of using R. These shells vary in the degree of statistical functionality and control the user has over statistical options. The examples and R scripts used in this book were all generated using R Studio. R Studio is a shell which adds a four-paneled organization to easily display the console, script files, packages, figures, and files. R Studio also makes it easy to read data into R by allowing a point and click interface for data imports. Other than data in and package loading, R Studio is still a syntax-based format so it maintains all the statistical and programming functionality of base R.

For the researcher who is more comfortable working with a fully point and click format, there are several R shells which run R through a point and click menu driven format. R Commander is a very basic R shell which adds point and click menus for data management and basic statistical functions but also allows for R scripts so complete R functionality is available through the use of syntax. Other shells such as JASP, Jamovi, or Rattle provide complete point and click formats for R making the experience similar to working with SPSS software. These shells allow for point and click menus for a wide selection of basic and more sophisticated statistical methods; however, depending on the shell, the syntax may not be available to manipulate so there will be less control over specific statistical options. Readers are encouraged to try out different shells for R to find the presentation that works best for their purposes.

R PACKAGES

R software comes segmented into a base program and a host of easily integrated packages containing specialized content. This is one of the most beneficial elements of R software as it allows the program to be current and comprehensive with users developing new statistical packages all the time, thus continually adding functionality into the program. Using the R console and GUI, users can download the necessary packages for the added functionality of interest (usually this is as simple as going to a dropdown menu or checkbox and then choosing a package to download). If you already know the name of the package to be installed, you can also install packages using the syntax `install.packages("package _ name")`. Then, to load the package, the syntax `library()` can be used to load the package into R so its functionality can be used. Multiple regression is a part of the R base software so no package loading is required to run this topic. Take, however, the example topic hierarchical linear modeling covered in Chapter 9 of this book. Hierarchical linear modeling is not a part of the base R package but the capability can be added into R by loading the {lme4} package. First the user must make sure the {lme4} package has been downloaded into R by choosing it from a check list or dropdown menu. Then the lme4 package can be installed by the syntax `library (lme4)`.

```
Console ~/
> library(lme4)
```

An image of the package installation interface in R Studio.

	Name	Description	Version	
☐	nms	Pretty Time of Day	0.4.2	⊗
☐	htmlTable	Advanced Tables for Markdown/HTML	1.13.1	⊗
☐	htmltools	Tools for HTML	0.3.6	⊗
☐	htmlwidgets	HTML Widgets for R	1.3	⊗
☐	jsonlite	A Robust, High Performance JSON Parser and Generator for R	1.6	⊗
☐	KernSmooth	Functions for Kernel Smoothing Supporting Wand & Jones (1995)	2.23–15	⊗
☐	knitr	A General-Purpose Package for Dynamic Report Generation in R	1.21	⊗
☐	labeling	Axis Labeling	0.3	⊗
☐	languageR	Analyzing Linguistic Data: A Practical Introduction to Statistics	1.5.0	⊗
☐	lattice	Trellis Graphics for R	0.20–38	⊗
☐	latticeExtra	Extra Graphical Utilities Based on Lattice	0.6–28	⊗
☐	lavaan	Latent Variable Analysis	0.6–3	⊗
☐	lazyeval	Lazy (Non-Standard) Evaluation	0.2.1	⊗
☑	lme4	Linear Mixed-Effects Models using 'Eigen' and S4	1.1–19	⊗
☐	lpSolve	Interface to 'Lp_solve' v. 5.5 to Solve Linear/Integer Programs	5.6.13	⊗
☐	magrittr	A Forward-Pipe Operator for R	1.5	⊗
☐	maps	Draw Geographical Maps	3.3.0	⊗
☐	maptools	Tools for Handling Spatial Objects	0.9–4	⊗
☐	markdown	'Markdown' Rendering for R	0.9	⊗
☐	MASS	Support Functions and Datasets for Venables and	7.3.51.1	

THE R CONSOLE AND SCRIPTS

R software is a completely syntax driven statistical program which is essentially a fully functional computer programming language with pre-programmed statistical functions. When you open up R (through any R shell) you start with a completely blank slate. You can type syntax directly into the R console and the output will appear directly below the syntax in the console window.

```
Console ~/Desktop/
> 5+5
[1] 10
>
```

Although typing syntax directly into the R console is straightforward, the R console is not easy to edit. Therefore, if more than just basic operations are desired, the R user may desire to work from R scripts. R scripts provide a basic text window where R code can easily be written and edited prior to running. When writing R scripts, comments can also be placed into the syntax to make the code more readable. Any text placed after a '#' in the syntax file will be ignored by R when run, thus allowing a way to easily annotate the file.

```
● ● ●                           RStudio Source Editor
Appendix_1.R ×
    ↶  ☐  ☐ Source on Save  Q  ⁄  ▾  ☐              → Run  ⇥  ☄ Source ▾  ☰
 1   #Example of script file with comments.
 2   #This syntax does the sum of 5 plus 5
 3
 4   5+5
 5
```

BASIC OPERATIONS IN R

R software can easily be used to do basic mathematical functions, create variables, and do basic data management.

```
● ● ●                           RStudio Source Editor
Appendix_2.R ×
    ↶  ☐  ☐ Source on Save  Q  ⁄  ▾  ☐              → Run  ⇥  ☄ Source ▾  ☰
 1▾  #### Simple Mathematical Operationst####
 2
 3   5 + 5     # Simple operations
 4
 5   10 - 5  # Subtraction
 6
 7   5*5       # Multiply
 8
 9   25/5      # Divide
10
11   5^2       # Exponents
12
13   5^3
14
```

```
● ● ●                           RStudio Source Editor
Appendix_3.R ×
    ↶  ☐  ☐ Source on Save  Q  ⁄  ▾  ☐              → Run  ⇥  ☄ Source ▾  ☰
 1▾  ####Naming Objects(Variables)####
 2
 3   x <- 5  # Naming objects
 4
 5   y <- 10
 6
 7   x          # Call an object to see its value
 8
 9   x + y   # Simple operations with objects
10
11   x - y
12
13   x*y
14
15   x/y
16
17▾  ###########################
18
19   z <- x*2  # Defining new objects with existing objects
20
21   w <- y/2
22
23   xy <- x*y
24
25   xy    # To check your result, simply call the object name
26
```

The procedure for naming objects and variables shown in script Appendix_A.R can be used for a variety of purposes. It can be used as shown as a way to define variables and then perform operations on those variables. It can also be used as a way to name statistical models to more easily keep track of which model is which when multiple models are run.

BASIC STATISTICAL FUNCTIONS

Pictured in the Appendix A script, you can find the R syntax for basic statistical functions (mean median mode, variance, etc.). Each of these functions can easily be computed by filling in the parenthesis () with the variable or vector name of interest. The Appendix A script creates two vectors for the interested reader to then practice computing these functions.

```
RStudio Source Editor

Appendix_4.R

                 Source on Save                          Run      Source

 1  x <- c(5, 1, 3, 5, 4, 2, 2, 5, 5, 3) # To create a vector, or list of values, use the concatenate function
 2  y <- c(1, 0, 1, 2, 3, 3, 4, 3, 3, 2)
 3
 4  mean(x)   # To see the mean of a vector, simply use the "mean()" function
 5  mean(y)
 6
 7  sd(x)      # To see the standard deviation ofa vector, use the "sd()" function
 8  sd(y)
 9
10  var(x)    # Similarly, to see the variance of a vector, use the "var()" function
11  var(y)
12
13  mode(x)   # To see the mode, use "mode()"
14  mode(y)
15
16  median(x) # To see the median, use "median()"
17  median(y)
18
19  max(x)     # To see the maximum value, use "max()"
20  max(y)
21
22  min(x)     # To see the minimum value, use "min()"
23  min(y)
24
25  range(x)  # To see the range, use "range()"
26  range(y)
27
28  log(x)     # To calculate the log of a vector, use "log()"
29  log(y)
30
31  sqrt(x)    # To calculate the square root of a vector, use "sqrt()"
32  sqrt(y)
```

The goal of this appendix was to provide a basic map for the navigation of the R environment. There is obviously so much more that the R platform can do. However, if the user understands the package-based structure and the use of R scripts and can feel comfortable using basic functions they should have the background to use this book or teach themselves any analysis they need to perform.

LOADING/IMPORTING DATA

Unless you are generating your own data within R, you will generally begin each R session by loading the data of interest into the R platform. This can be done through various syntax functions within R or, depending on the R shell used, sometimes through a point and click process.

One very useful feature of R is that it can import data from a variety of formats.

```
-Text file: data<- read.table("c:/file.csv", header=TRUE)
-Excel file: data<- read.excel("c:/file.xls",
sheetName="sheet1")
-SPSS file: data<- library(foreign) data<- read,spss("c:/
data.sav")
-SAS file: data<- library(haven) data<- read_sas("c:/
data.sas7bdat")
```

As mentioned previously, if you are using an R shell like R Studio or R Commander, there are options to simply point and click to read in data. This textbook has used R Studio to run R. In R Studio, there is an option in the upper right-hand quadrant that allows the user to import data. The user can choose from text, Excel, SPSS, SAS, or Stata formats.

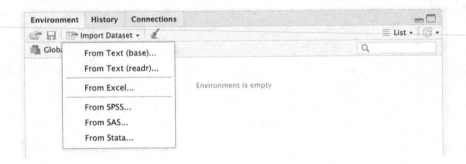

From here the user is sent to a window where they can select and rename their file prior to reading it into R. As the data are selected, the code preview at the bottom of the box shows the actual R code that will be used to read in the file. Here, the user can manipulate the code if different options are desired.

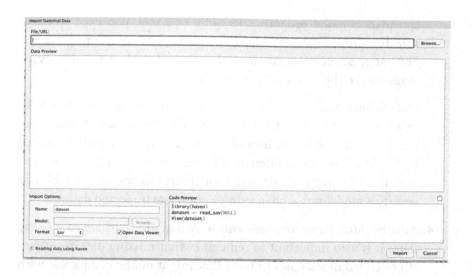

TROUBLESHOOTING: WHY WON'T MY SYNTAX RUN?

Every R user will run into times when the code just won't run. First and foremost, don't give up! This does not mean you don't know what you are doing. This happens to all R users, not just users learning the basics. If you are having difficulty troubleshooting your problem, here are a few strategies to run through.

1. *'Look for the dumb'*. I have taught graduate level statistics courses using R software for 13 years and this is always how I describe this to my students. When you are first troubleshooting a syntax problem, don't assume you did it wrong. Look for the little mistakes first. Did you spell all the code right? Did you forget a parenthesis or quotation mark? Are your variable names correct (capitalization/lower case does matter)? With my students, I find that nearly 99% of the time, the problem is something incredibly tiny.

2. *Are all necessities loaded?* If everything you have written appears to be spelled properly and all punctuation and syntax appear to be correct to the best of your knowledge, the next place to look would be to make sure all necessities are loaded. As mentioned previously, R is a package-based program. If you are running an analysis that is not part of the base R package, you will need to have the proper R package loaded before running your analysis. Similarly, since R is not a

spreadsheet-based program like SPSS or Excel, you need to make sure the proper dataset is loaded into R prior to running syntax. Thus suggestion #2: check to make sure all packages and relevant datasets/variables are properly loaded.

3. *Data format.* Sometimes certain R functions are expecting data to be in a very specific format. Simply using the entire dataset to run the function may not be the format required. Check the format needed by the function (this is often detailed in the errors you receive). You may need to create a data frame or matrix of specific variables or specify a model rather than using a complete imported dataset.

4. *Missing data.* Some analyses and R syntax are sensitive to missing data. It is also important to remember that missing data are sometimes coded in a dataset in ways that might not be compatible with the statistical software you are using (for example, if it is coded as NA or 999). If your code still is not working, check to see if you have missing data and how it is coded in your dataset. You may need to do some data cleaning or missing data removal/replacement prior to analysis.

5. *Debugging: process of elimination.* If you have now ruled out silly mistakes and made sure all necessities and missing data are dealt with, it is time to figure out where the problem in the code is. There are many approaches to debugging but only one simple approach will be described here: the process of elimination. If your line of code is not working, one excellent way to find the problem is to strip the code down to its most basic form and test each element to find where the issue is. For example, if your line of code has multiple predictor variables and several customization options, first strip the code down to the most basic model with just one predictor and no customizations. Run this line of code. If this does not work, there is likely a problem with the function call or the package. If this does work, add in another variable. If this works, add in another variable. If this works, add in a customization option. Working in this manner you can pinpoint exactly which piece of the model is causing the problem.

R HELP

This appendix has sought to provide basic R for the users of this text. If you desire further support for R software, there are many published texts which may be useful. A list will be provided at the end of this appendix. Quick help on functions can also be found by using the ?() function in R. For example, if you type ?lm into the R console, the help file for the lm() function will be called up. These same help files can be found on the R Project website (R-project.org) as well as in the bottom right quadrant in R Studio. There are also many different R help options on the internet such as search.r-project.org and Rseek.org.

R RESOURCES

Crawley, M. (2013). *The R book*. Wiley & Sons Ltd.

Field, A., & Miles, J. (2012). *Discovering statistics using R*. Sage Publications.

Appendix B: Non-Parametric Analysis Based on Ranks

CHAPTER 2 DISCUSSES TWO different non-parametric measures of relationship: the Spearman rho and the Kendall tau. Each of these is a non-parametric statistical analysis based on rank ordered data, rather than raw data. This appendix will briefly demonstrate the process of rank ordering data so the interested reader can better understand why the non-parametric correlations may be better for data with assumption violations.

The Process of Ranking Data

Variable 1
23.7
24.7
26.7
21.1
22.5
25.1
23.2
21.8
22.1
25.1
22.1

Rank ordering data is exactly what it sounds like. We will rank order the cases from lowest to highest. Here the lowest value is 21.1 so it will receive a rank of 1. Next is 21.8 so it will receive a rank of 2. Continue this process until the last variable is reached.

Sometimes, as is the case with this example, ties occur. Here we have two sets of ties: two people with a score of 22.1 and two people with a score of 25.1. The most common way to deal with this is to give each the average of the rank places they take up (so for example, 22.1 would be taking up spaces 3 and 4: the average would be 3.5) then the next case would have the next rank up.

ID	Variable 1	Rank V1
1	23.7	7
2	24.7	8
3	26.7	11
4	21.1	1
4	22.5	5
6	25.1	9.5
7	23.2	6
8	21.8	2
9	22.1	3.5
10	25.1	9.5
11	22.1	3.5

This is the basic process! It is important to note that different rank-based analysis will work with this process slightly differently. For example, for Spearman rho, we would go through this ranking process for each variable to be correlated and then just use the Pearson R formula on the ranks. This is similar to the process for Kendall tau except for tau, once data are ranked, rank inversions are computed and a different formula is used for computation. This process is also used for rank-based means comparison procedures with the exception being ranking is done across all groups rather than within variables so that group comparisons can be made.

EXAMPLE: RANKING FOR COMPUTATION OF SPEARMAN RHO AND KENDAL TAU: RANKING WITHIN VARIABLE

ID	Variable 1	Rank VI	Variable 2	Rank V2
1	23.7	7	30.5	10
2	24.7	8	30.1	9
3	26.7	11	28.3	7
4	21.6	1	23.6	1
5	22.5	5	26.8	3
6	25.1	9.5	27.1	4
7	23.2	6	27.7	5
8	21.8	2	26.7	2
9	22.1	3.5	28.2	6
10	25.1	9.5	28.4	8
11	22.1	3.5	31.3	11

EXAMPLE: RANKING FOR COMPUTATION OF NON-PARAMETRIC MEANS COMPARISONS: RANKING ACROSS ALL GROUPS

Group 1	Rank VI	Group 2	Rank V2
61.6	15	56.1	12
58.9	13	63.9	16
41.6	2	53.3	9
64.4	17	61.5	14
54.3	11	44.9	3
32.3	1	69.1	18
47.4	5	47.7	6
46.3	4	47.8	7
48.1	8	53.4	10

Appendix C: R Function and Package Index

THIS INDEX CATALOGS THE R functions and packages used in this text. This can be used as a quick way to look up the necessary package to install if you have the function but don't remember the package.

{Package}

Function()

BASIC FUNCTIONS

{base}	#this package is usually already installed in R
AIC()	#returns the AIC value of a model
as.factor()	#designates a variable as a factor/categorical
as.matrix()	#creates a matrix out of identified columns of data
BIC()	#returns the BIC value of a model
ifelse()	#programming if/else statements. Can dummy code
log()	#computes the base 10 logarithm
mean()	#calculates the mean of a variable
sqrt()	#computes the square root
{graphics}	#this package is usually already installed in R
abline()	#places a regression line on scatterplot
hist()	#creates a histogram

plot()	#creates a scatterplot
qqnorm()	#creates a Q–Q plot

{stats}   **#this package is usually already installed in R**

anova()	#computes ANOVA or deviance comparison of fitted models
aov()	#runs an analysis of variance
cor()	#runs a single correlation
cor.test()	#runs a single correlation with significance test
lm()	#runs a single/multiple OLS regression
pairwise.t.test()	#runs unadjusted pairwise comparisons
predict()	#returns the predicted values from a fitted model
resid()	#returns the regression model residuals
step()	#runs stepwise regression
summary()	#returns detailed summary of function (type lm or lmer)
TukeyHSD()	#runs a Tukey pairwise comparison test

{moments}

kurtosis()	#returns the kurtosis statistic for a variable
skewness()	#returns the skewness statistic for a variable

CORRELATION AND REGRESSION

{Hmisc}

rcorr()	#runs a correlation matrix

{lm.beta}

lm.beta()	#returns regression output with standardized beta column

{Performance Analytics}

    chart.Correlation()   #runs a correlation matrix with scatterplots

ASSUMPTIONS

{lmtest}

    dwtest()   #calculates the Durbin Watson statistic

{olsrr}

    ols_col_diag()   #returns detailed multicollinearity statistics

INTERACTIONS

{bda}

    sobeltest()   #runs Sobel test of indirect effect

{interactions}

    johnson_neyman()   #returns Johnson Neyman output and plot for an interaction

{sjplot}

    plot.model()   #creates a plot of an interaction effect

REGRESSION EXTENSIONS

{caret}

    R2()   #computes cross-validation R squared

{fastdummies}

    dummy_cols   #creates dummy variables from same root name

{lme4}

    lmer()   #runs hierarchical linear models

{lsr}

etaSquared()   #runs the eta squared effect size for ANOVA models

{psych}

mediate()   #runs test of total, direct, and indirect effects

Appendix D: End of Chapter Exercise Script File Solutions

THESE SCRIPT FILES ALSO appear on the companion website. If desired, once the data set for each script is loaded, the file can be run in its entirety and full output for each question will be generated.

CHAPTER 2: END OF CHAPTER EXERCISES SCRIPT

```
##Chapter 2 Correlation Exercises
## load data chapter2ex.sav
##1. This code runs the correlaions and significance
tests and a scatterplot
cor(chapter2ex$effort, chapter2ex$performance)
cor.test(chapter2ex$effort, chapter2ex$performance)
plot(chapter2ex$effort, chapter2ex$performance)
#The correlation r=.61 p<.001 is positive, moderate and
significant.
#As effort increases, performance increases.
##2. This code runs Spearman Rho and Kendall Tau
cor(chapter2ex$distraction, chapter2ex$performance,
method="spearman")
cor.test(chapter2ex$distraction, chapter2ex$performance,
method="spearman")
cor.test(chapter2ex$distraction, chapter2ex$performance,
method="kendall")
#Spearman Rho and Kendall Tau are both negative,
significant and weak to moderate
#(rho=-.39, Tau=-.27). As effort distraction increases,
performance decreases.
```

```
#Kendall Tau may be a preferable choice of method given
the larger sample size.
#Spearman Rho is only exact for smaller samples.
```

CHAPTER 3: END OF CHAPTER EXERCISES SCRIPT

```
##Chapter 3 Regression Exercises
##load data chapter3ex.sav
##1. runs a simple linear regression model predicting
demand from advertising budget
ch3SR<- lm(demand~advertbudget, chapter3ex)
summary(ch3SR)
#This model is significant (F=124.9 p<.001) and accounts
for 25% of the variance #in demand. The intercept is
significantly different from zero. The slope is also
#significantly different from zero (b=14.186 p<.001)
indicating that advertising #budget is a significantl
predictor of demand.
#Regression Equation: predicted
demand=10803+14.186(advertbudget)
##2. runs a multiple regression model predicting demand
from price, availability and #advertising budget
library(lm.beta)
ch3MR<- lm(demand~price+availability+advertbudget,
chapter3ex)
summary(ch3MR)
summary(lm.beta(ch3MR))
#This model is significant (F=120.7, p<.001) and accounts
for 49% of the variance #in demand. The intercept is
significantly different from zero. Slopes for price,
#availability and advertising budget are all significant
and positive. Comparing #standardized betas, advertising
budget and availability are the strongest predictors #of
demand.
```

CHAPTER 4: END OF CHAPTER EXERCISES SCRIPT

```
##Chapter 4 Regression Exercises:
##load data chapter4ex.sav
##This code runs a multiple regression model predicting
demand from price, #availability and advertising budget
library(Hmisc)
library(PerformanceAnalytics)
library(lm.beta)
ch4MR<- lm(demand~price+availability+advertbudget,
chapter4ex)
```

```
summary(ch4MR)
summary(lm.beta(ch4MR))
##1. This code will run a matrix scatterplot to check
linearity
chapter4ex.matrix<- as.matrix(chapter4ex)
ch4_scatterplot<- chart.Correlation(chapter4ex.matrix)
#The scatterplots show no evidence of non-linear
relationships between predictors
#and outcome. The assumption of linearity is met.
##2. This code will create model residuals then run a
histogram and qq plot of #residuals
ch4.residuals<-resid(ch4MR)
hist(ch4.residuals)
qqnorm(ch4.residuals)
plot(ch4.residuals)
#The histogram of residuals looks normally
distributed. The qq plot of residuals #shows the residuals
falling along a 45 degree angle (indicating normality)
and the #residuals scatterplot shows the majority of
residuals near y=0 with fewer larger #residuals. The
assumption of normality is supported.
##3. Looking at the residuals scatterplot, there is
nearly equal variance in the #residuals for all predicted
values (the X axis). The assumption of homoscedasticity
#is supported. There are no visible patterns in the
residuals. The assumption of #independence is supported.
#4 This code will run the Durbin Watson statistic for
independence of errors
library(lmtest)
dwtest(ch4MR)
#The Durbin Watson is almost exactly 2. There is no
autocorrelation of residuals
#(independence assumption is supported)
#5. This code will run multicollinearity diagnostics
library(olsrr)
ols_coll_diag(ch4MR)
#The tolerance for all predictors is very high (the VIF
is very low) and the variance
#proportions show only one variable loading per
eigenvalue. There is an absence of
#multicollinearity.
```

CHAPTER 5: END OF CHAPTER EXERCISES SCRIPT

```
###Chapter 5 Regression Exercises
##Load data chapter5ex.sav
```

```
##1. This code will dummy code gender into variable "boy"
with 1=boy and 0=girl
#chapter5ex$boy<- ifelse(chapter5ex$gender==2, 1, 0)
##2. This code will dummy code grade into two dummy
variables "elementary" and #"middle"
chapter5ex$elementary<-ifelse(chapter5ex$grade==1, 1, 0)
chapter5ex$middle<-ifelse(chapter5ex$grade==2, 1, 0)
##3. prep work. This code will mean center age and
screentime.
##We need to mean center for the interaction.
chapter5ex$cage<- chapter5ex$age-mean(chapter5ex$age)
chapter5ex$cscreentime<- chapter5ex$screentime
-mean(chapter5ex$screentime)
##3. This code will run a regression model predicting
attention from screentime, age #and boy.
ch5full<- lm(attention~cscreentime+cage+boy, chapter5ex)
summary(lm.beta(ch5full))
##This code will add the interaction of screentime and
age to the previous model.
ch5int<- lm(attention~cscreentime+cage+boy+cscreentime
*cage, chapter5ex)
summary(lm.beta(ch5int))
##This code will plot the interaction of screentime and
age then run a Johnson Neyman #interval
library(sjPlot)
plot_model(ch5int, type="pred", terms=c("cscreentime",
"cage"))
library(interactions)
johnson_neyman(ch5int, cscreentime, cage, vmat=NULL,
alpha=0.05,
      plot=TRUE, title="Johnson-Neyman plot")
#This model is significant (F=37.46 p<.001) and accounts
for 52% of the variance #in screentime. Screentime and
age are both significant negative predictors of
#attention. As screentime and age increase, attention
decreases. Gender and the #interaction of screentime and
age were not significant predictors of attention.
```

CHAPTER 7: END OF CHAPTER EXERCISES SCRIPT

```
#####Chapter 7 Exercises: Hierarchical Regression
##load data chapter7ex.sav
##1 pre-work. This code will run a regression model
predicting attention from only an #intercept.
interceptonly<-lm(attention~1, chapter7ex)
##1a. This code will run a regression model predicting
attention from screentime, age #and boy
```

```
ch7model1<- lm(attention~screentime+age+boy, chapter7ex)
summary(ch7model1)
```
##1b. This code will run a regression model predicting attention from screentime, age, #boy and the interaction of screentime and age.
```
ch7model2<- lm(attention~screentime+age+boy+screentime*age, chapter7ex)
summary(ch7model2)
```
##1c. This code will compare models 1a and 1b.
```
anova(ch7model1, ch7model2)
```
##2a. This code will run a forward regression predicting attention from screentime, #age and boy.
```
ch7forward<-step(interceptonly, scope=formula(ch7model1), direction="forward")
summary(ch7forward)
ch7forward$anova
```
##2b. This code will run a backward regression predicting attention from screentime, #age and boy.
```
ch7backward<-step(ch7model1, direction="backward")
summary(ch7backward)
ch7backward$anova
```
##3a. This code will run a regression predicting attention from screentime and age
```
ch7modelA<- lm(attention~screentime+age, chapter7ex)
summary(ch7modelB)
```
#3b. This code will run a regression predicting attention from screentime and boy.
```
ch7modelB<- lm(attention~screentime+boy, chapter7ex)
summary(ch7modelB)
```
##3c. THis code will use AIC and BIC stats to compare models 3a and 3b.
```
AIC(ch7modelA)
AIC(ch7modelB)
BIC(ch7modelA)
BIC(ch7modelB)
```
##Model 3a is the better fit model because the AIC and BIC stats are smaller than the #AIC and BIC stats for model 3b.

CHAPTER 8: END OF CHAPTER EXERCISES SCRIPT

###Chapter 8 Exercises - moderation, mediation, regression discontinuity
##Load data chapter8ex1, chapter8ex2, chapter8ex3
##1. Moderation
##This code will dummy code gender into "boy where 1=boy and=girl

```
chapter8ex1$boy<- ifelse(chapter8ex1$gender==2, 1, 0)
##This code will create centered predictors for the
interaction we are about to use
chapter8ex1$cboy<- chapter8ex1$boy-mean(chapter8ex1$boy)
chapter8ex1$cage<- chapter8ex1$age-mean(chapter8ex1$age)
## This code will run a model predicting screentime from
age, boy and the interaction #of age and boy. This
interaction tests whether gender is a moderator between
age and #screentime.
ch8mod<- lm(attention~screentime+cage+cboy+cage*cboy,
chapter8ex1)
summary(ch8mod)
plot_model(ch8mod, type="pred", terms=c("cage", "cboy"))
library(interactions)
johnson_neyman(ch8mod, cage, cboy, vmat=NULL, alpha=0.05,
      plot=TRUE, title="Johnson-Neyman plot")
```

##This model is significant (F=39.57 p<.001) and accounts for 53% of the variance #in attention. Screentime, age and boy are all significant negative predictors of #attention. as screentime increases, age increases and gender is male, attention is #lower. The interaction between boy and age is not significant indicating that gender #is not a moderator of this relationship. The relationship between age and screentime #is the same regardless of gender.

2. Regression Discontinuity Data

##This code will run a simple regression discontinuity model modeling and plotting
#intercept differences but no slope differences

```
ch8rd<-(lm(foodneeds~householdincome+noassistance,
chapter8ex2))
summary(ch8rd)
chapter8ex2$predsimple<-predict(ch8rd)
plot(chapter8ex2$householdincome,chapter8ex2$foodneeds)
with(subset(chapter8ex2, noassistance==1),lines(household
income, predsimple, type="l", lwd="3"))
with(subset(chapter8ex2, noassistance==0),lines(household
income, predsimple))
```

##This model is significant (F=535 p<.001) accounting for 87% of the variance in
#foodneeds. Income and the dummy variable for noassistance were both significant.
#The higher the income, the lower the food needs. Those without assistance had #significantly higher food needs

3. Mediation

##This code will create a correlation matrix.

```
library(Hmisc)
```

```
chapter8ex3_corr<- as.matrix(chapter8ex3[c(1, 2, 3)])
rcorr(chapter8ex3_corr)
#Relationships between the outcome, mediator and
predictor are all signficant and #moderate to strong
satisfying Baron and Kenny's steps 1-3.
##This code will run two regression models. One predicting
behavior from the #predictor, age. The other predicting
behavior from the predictor age and the proposed
#mediator
library(lm.beta)
ch8main<- lm.beta(lm(behavior~age, chapter8ex3))
summary(ch8main)
ch8med<- lm.beta(lm(behavior~age+selfcontrol,
chapter8ex3))
summary(ch8med)
anova(ch8main,ch8med)
##Once the mediator, self control, was added to the
model, the effect of age is still
#significant but its contribution is greatly reduced. This
is evidence of a partially
#mediated model.
#This code will run the Sobel test for significance of
the indirect effect
library(bda)
mediation.test(chapter8ex3$selfcontrol, chapter8ex3$age,
chapter8ex3$behavior)
#This code will run total, direct and indirect effects as
well as obtain bootstrapped
#confidence intervals for the indirect effect.
library(psych)
mediate(y=behavior~age+(selfcontrol), data=chapter8ex3)
#The total effect of age on behavior is significant. The
direct effect of age on #behavior is also significant. The
indirect effect is significant (see Sobel result #or
bootstrapped CI) indicating the effect of the mediator is
significant. This is #evidence of partial mediation.
```

CHAPTER 9: END OF CHAPTER EXERCISES SCRIPT

```
####Chapter 9 Exercises
##load data chapter9_train, chapter9_cross
##1. This code will plot the relationship between age and
expense
plot(chapter9_train$age, chapter9_train$expense)
#2.This code will compute log and square root
transformations to linearize the #relationship
log_age<-log(3+chapter9_train$age)
```

```
sqrt_age<- sqrt(3+chapter9_train$age)
log_expense<-log(150+chapter9_train$expense)
chapter9_train$sqrt_expense<- sqrt(150+chapter9_
train$expense)
#This code will plot the different log/square root
options for linearizing.
plot(log_age, log_expense)
plot(log_age, chapter9_train$expense)
plot(chapter9_train$age, log_expense)
plot(sqrt_age, chapter9_train$sqrt_expense)
plot(sqrt_age, chapter9_train$expense)
plot(chapter9_train$age, chapter9_train$sqrt_expense)
#This code will run the model predicting square root of
expense from age and male.
sqrtmodel<- lm(sqrt_expense~age+male, chapter9_train)
summary(sqrtmodel)
#This model is significant (F=36.81, p<.001) and accounts
for 12% of the variance #in expense. Age is not a
significant predictor but male is (being male - higher
#expense)
#3. This code creates a non linear term for age.
chapter9_train$age2<- chapter9_train$age*chapter9_
train$age
#This code runs a regression predicting expense from age,
age squared and gender.
trainingmodel<-lm(expense~age+age2+male, chapter9_train)
summary(trainingmodel)
#This model is significant (F=486.6 p<.001) and accounts
for 74% of the variance in #expense. The quadratic term
for age and gender are both significant.
#4 This code begins the cross validation for the model in
#3. First predicted values are obtained.
predictions<- 3.8234+(-.6606*chapter9_cross$age)+(-10.
5419*chapter9_cross$age*chapter9_cross$age)+
  (16.9203*chapter9_cross*male)
#This code computes the cross validation R squared
library(caret)
R_squared=R2(predictions,chapter9_cross$expense)
#This code runs the model on the cross validation sample
to assess coefficient #stability
chapter9_cross$age2<- chapter9_cross$age*chapter9_
cross$age
crossmodel<-lm(expense~age+age2+male, chapter9_cross)
summary(crossmodel)
#The coefficients are in the same direction and relative
strength for both the #training and the cross validation
models indicating model stability.
```

CHAPTER 10: END OF CHAPTER EXERCISE SCRIPT

```
####Chapter 10 Exercises
## load data chapter10ex1, chapter10ex2
###1. This code will dummy code school for use in the FEM
library(fastDummies)
ch10_data<- dummy_cols(chapter10ex1,
select_columns="school")
##This code will run the fixed effects model predicting
getotal from npamem, geread #age and school
ch10_FEM<- lm(getotal~npamem+gereadcm+age+sch
ool_1+school_2+school_3+school_4+school_5+school_6+
        school_7+school_8+school_9+school_10, ch10_data)
summary(ch10_FEM)
ch10_basic<- lm(getotal~npamem+gereadcm+age, ch10_data)
summary(ch10_basic)
anova(ch10_basic, ch10_FEM)
##This model is significant (F=105.3 p<.001) and accounts
for 70% of the variance in #getotal. npamem and age are
significant predictors of getotal. The higher the #reading
and the lower the age, the better the getotal. Accounting
for school level #variance significantly improved
variance explained beyond just the predictors alone.
library(lme4)
###load data chapter10ex2
##2. This code will run a null HLM model predicting
getotal from an intercept.
ch10null<- lmer(getotal~1+(1|classroom), chapter10ex2)
summary(ch10null)
##3. This code will run a random intercepts model
predicting getotal from npamem, #gereadcm and ptratio
ch10RI<- lmer(getotal~npamem+gereadcm+ptratio+(1|class
room), chapter10ex2)
summary(ch10RI)
##4. This code will run a random coefficients model
predicting getotal from npamem, #gereadcm and ptratio
with gereadcm allowed to vary across classrooms.
ch10RC<- lmer(getotal~npamem+gereadcm+ptratio+(gereadc
m|classroom), chapter10ex2)
summary(ch10RC)
##npamem and gereadcm are significant positive predictors
of getotal.
```

Appendix E: Glossary

TERM (CHAPTER FIRST APPEARS IN)

Akaike information criterion (AIC) (Chapter 7): general model comparison statistic. Can compare nested or non-nested models.

Analysis of variance (ANOVA) (Chapter 5): statistical analysis partitioning total variance into variance explained by the independent variables and variance left unexplained.

Backward deletion (Chapter 7): stepwise model selection criteria which begin with all eligible predictors in the regression model then delete predictors one at a time which do not significantly contribute to the R squared.

Bayesian information criterion (Chapter 7): general model comparison statistic. Can compare nested or non-nested models.

Beta or standardized beta (Chapter 3): standardized regression slope coefficient. Can be used to compare relative strength of predictor variables.

Bootstrapped confidence intervals (Chapter 8): method for testing significance of the indirect effect in mediation analysis that does not assume normally distributed sampling distribution.

Centering (Chapter 5): creating a new variable equal to the variable minus the mean of the variable. Centering predictors can help reduce multicollinearity issues introduced by using interaction effects.

Condition index (Chapter 4): measure of multicollinearity. Often considered problematic when greater than 30.

Correlation (Pearson product moment correlation coefficient) (Chapter 2): quantification of the relationship between two variables. Ranges from –1 to 1. Can be interpreted in terms of strength and direction and has an associated significance test to inform if the relationship is present in the population.

Covariance/covariation (Chapter 2): the most basic quantification of the relationship between two variables. When variables used are standardized, this measure is equivalent to the Pearson correlation.

Cross-level interaction (Chapter 10): in a hierarchical linear model when there is an interaction between an individual level variable and a context level variable.

Cross-validation (Chapter 9): the process of validating a multiple regression model on a second sample from the same population to check for model stability and generalizability.

Direct effect (Chapter 8): mediation terminology. The effect of the predictor (X) on the outcome (Y) after controlling for the mediator.

Dummy variables (Chapter 5): recoding of categorical variables to a 0/1 scale for use in regression analysis.

Durbin Watson test (Chapter 4): test for autocorrelation of residuals. Provides evidence for the assumption of independence in regression.

Explanatory variable (Chapter 3): also called independent variable, predictor variable or regressor. These are the variables (Xs) explaining the outcome variable in a regression model.

Fixed effects (Chapter 10): the main effects of each predictor variable in a hierarchical linear model.

Fixed effects modeling (Chapter 10): use of dummy variables in a regression model to account for context level variance in the presence of nested levels of influence.

Forward regression (Chapter 7): method of stepwise model selection which enters predictors into a regression model one at a time based on the strongest increment in R squared.

Hierarchical linear modeling (Chapter 10): regression-based method for analyzing nested levels of influence allowing for the intercept and regression slopes to randomly vary across context.

Histogram (Chapter 4): graph to visualize distributions. Plots the frequency of each score in a distribution. Useful in determining if a variable is normally distributed.

Homoscedasticity (Chapter 4): assumption of multiple regression. States that there must be equal variance in the prediction errors (residuals) for all predicted values.

Independence (Chapter 4): assumption of multiple regression. States that the prediction errors (residuals) must be uncorrelated.

Indirect effect (Chapter 8): mediation terminology. Effect of the predictor *X* on the outcome *Y* through the mediator (*M*).

Interaction (Chapter 5): when the effect of one variable on the outcome depends on the level of another variable. Interactions are mathematically defined as the product of the two interacting variables.

Johnson Neyman interval (Chapter 5): analysis providing detailed information regarding how the slope coefficient changes when the interaction is significant.

Kendall tau (Chapter 2): a non-parametric rank-based correlation useful when assumptions of Pearson correlation are not met.

Kurtosis (Chapter 4): refers to proper height and spread of a distribution.

Linearity (Chapter 2): refers to the shape of the relationship between two variables. Pearson correlation and OLS regression assume the shape of this relationship to be linear, or able to be described with a straight line.

Mediation (Chapter 8): when a third variable explains the relationship between the predictor and the outcome. Can be tested using hierarchical regression and bootstrapped confidence intervals.

Model fit (Chapter 3): a set of statistics that quantify the overall strength or explanatory power of the regression model.

Moderation (Chapter 8): when a third variable changes the relationship between the predictor and the outcome. Can be tested using interactions.

Multicollinearity (Chapter 4): term for predictors accounting for overlapping variance in the outcome variable. When predictors are highly correlated with one another they will account for overlapping variance in the outcome variable leading to predictor redundancy and in extreme cases non-interpretable or nonsensical results.

Multiple R (Chapter 3): a measure of model fit. The correlation between the predicted outcome values and the actual outcome values.

Multiple regression analysis (Chapter 3): statistical analysis predicting one continuous outcome from one or more continuous explanatory (predictor) variables.

Multivariate normality (Chapter 4): assumption of multiple regression. States that the predictor variables and all linear combinations of the predictor variables must be normally distributed.

Nested models (Chapter 7): when regression models are systematically created such that models to be compared can be obtained by simply adding or subtracting predictors. In other words, Model 2 can be obtained by adding predictor(s) to Model 1.

Nested data (Chapter 10): when the cases in your sample are situated within a larger context with multiple levels. Classic example is a sample of students nested within different classrooms.

Non-linear term (Chapter 9): a polynomial term incorporated into a regression model in order to model non-linear relationships between variables.

Normality of residuals (Chapter 4): assumption of multiple regression. States that the errors in prediction (residuals) must be normally distributed.

Negative relationship (Chapter 2): the relationship between two variables has a negative coefficient indicating when the value of one variable increases, the value of the other variable tends to decrease.

Null model (Chapter 10): simplest hierarchical linear model. Predicts the outcome from just a random intercept.

Ordinary least squares (OLS) criterion (Chapter 3): this is the criterion many traditional regression models use to determine the best fitting model. OLS defines the best fitting model to be the model which minimizes the sum of squared deviations between the predicted values and the actual values.

Outcome variable (Chapter 3): also called dependent variable or regressand. This is the variable (Y) being explained or predicted by the explanatory (predictor) variables in the regression model.

Outlier (Chapter 4): a case very different from the rest of the sample potentially due to non-equivalent circumstances or error.

Overfitting (Chapter 9): when a regression model is fit too closely to characteristics of the sample such that the resulting model is a poor fit for the population.

Positive relationship (Chapter 2): the relationship between two variables has a positive coefficient indicating when the value of one variable increases, the value of the other variable also tends to increase.

Quantile-quantile (Q–Q) plot (Chapter 4): graph visualizing variable distribution. Plots the observed data against what the data would look like if they were normally distributed. Useful in determining if a variable is normally distributed.

Random effects (Chapter 10): the error terms in a hierarchical linear model. At individual level, this is equivalent to prediction error.

At the context level this represents the amount the intercept and slope(s) vary across contexts.

Random coefficients model (Chapter 10): a hierarchical linear model allowing both the intercept and at least one slope coefficient to vary across contexts.

Random intercepts model (Chapter 10): a hierarchical linear model allowing the intercept to vary across contexts but all slope coefficients are fixed effects.

Reference category (Chapter 5): The '0' category when a variable is dummy coded. This will be the category each dummy variable category coded '1' will be compared to.

Regression discontinuity (Chapter 7): regression application using dummy variables and interactions to answer research questions about treatment/control situations. In these situations, assignment to the 'treatment' and 'control' group is determined based on a cut point on a continuous predictor.

Residual (Chapter 3): predicted values obtained from the regression prediction equation minus the actual values of the outcome variable. A measure of prediction error.

Residuals plot (Chapter 4): plot used to check regression assumptions. Plots standardized prediction errors (residuals) on the Y axis and predicted values of the outcome on the X axis.

Restriction of range (Chapter 2): when variables used in a correlation or regression are truncated such that the variable does not fully represent the full range of values. Interpretations must be made carefully with the knowledge that the relationship may be underestimated.

R squared (Chapter 3): a measure of model fit. The proportion of variance in the outcome variable explained by the predictor variables in the regression model.

Sequential regression (Chapter 7): comparison of nested models where predictor order of entry is determined by theory and discretion of the researcher.

Scale of measurement (Chapter 2): the way in which a variable is measured. Most commonly in reference to the characteristics of the real number line that a variable has. For correlation and regression, it is important that variables used be continuous (interval or ratio scales of measurement) or if categorical (nominal or interval) treated carefully.

Scatterplot (Chapter 2): a plot of the relationship between two continuous variables with one variable on the X axis and the other variable on the Y axis. Can be used to visually represent relationship between two variables.

Simple linear regression (Chapter 3): statistical analysis predicting one continuous outcome from one continuous explanatory(predictor) variable.

Skewnness (Chapter 4): refers to asymmetry of a distribution.

Slope coefficient (Chapter 3): quantifies the degree the outcome changes as the predictor changes. The steepness of the regression line. Represents the relative contribution of the predictor to the outcome.

Sobel test (Chapter 8): test of significance of the indirect effect in a mediation analysis.

Spearman rho (Chapter 2): a non-parametric rank-based correlation equivalent to the Pearson correlation ·computed on ranks instead of raw data. Most accurate in smaller samples.

Standard error of the estimate (Chapter 3): a measure of model fit. The average amount of error made when predicting the outcome variable from the predictor variables in the regression model. Always in the metric of the outcome variable.

Stepwise regression (Chapter 7): a class of nested model comparisons where the computer mathematically determines the order of entry or removal of predictors based on internal rules.

Tau matrix (Chapter 10): matrix of the variances and covariances of the random effects in a hierarchical linear model.

Transformation (Chapter 9): a mathematical function applied to a variable which cosmetically changes the distribution shape while maintaining cases in their same relative position.

Tolerance (Chapter 4): measure of multicollinearity. 1/VIF. Defined as the variance not accounted for when each predictor is predicted by the other predictors.

Total effect (Chapter 8): mediation terminology for the effect of the predictor (X) on the outcome (Y) without partialling out the mediator.

Variance (Chapter 2): quantification of the variability within a sample. Mathematically defined as the average squared deviation scores from the mean. Can be decomposed into variance explained by the variables in the model and variance left unexplained.

Variance proportions (Chapter 4): measure of multicollinearity. Uses eigenvalues to show how variance is proportioned across predictors.

Variance inflation factor (Chapter 4): measure of multicollinearity. Values larger than 10 considered problematic.

Y-intercept (Chapter 3): in a regression model, the value of Y when $X = 0$. The predicted value of Y when X is equal to zero.

Index

A

Academic dishonesty, 2
Academic test anxiety, 1
Adjusted R^2, 31–32
Akaike information criterion (AIC), 82
Analysis of covariance (ANCOVA), 76
Analysis of variance (ANOVA), 69–71
    as regression, 71–75
        equivalence, 75–77
ANCOVA, *see* Analysis of covariance
ANOVA, *see* Analysis of variance
Assumption of independence
    multiple regression analysis, 42
    Pearson *r*, 14–15

B

Backward deletion, 85
    R example, 89
Bayesian information criterion (BIC), 82
Bootstrapped confidence intervals, 106

C

Categorical variables, 55–57
Centering predictors, 65–67
Cheating dataset, 35
Cheating.sav dataset, 20
Condition index, 52
Correlation analysis, 3, 7, 23
    covariation, 9–10
    linear relationships, 10–11
    visualizing relationships, 7–9
Correlation interpretation, test anxiety
    data, 11–14

Correlation using R, 20
    {Hmisc} package, 21–22
    {Performance Analytics} package, 22–23
    {stats} package, 20–21
Covariation, 9–10
Cross-level interactions, 133
Cross-validation, 120–121
    coefficient stability example, 123–125
    procedures, 121–122
    R cross-validation R squared, 122–123
    samples, 121

D

Debugging, 148
Discordant pairs, 18
Dummy variables, 57
    face mask usage, 58
    in regression model, 58–60
    with 3+ levels, 60–61
    with two levels, 60
    using R, 60
    vision impairment, 57
    '0 0' category, 58
Durbin Watson test, 50
    R printout, 51

E

Eigenvalues, 52

F

Fixed effects modeling, 127, 128, 133, 138
    R using PT_FEM data, 128–130
Forward regression, 85

G

General linear model, 70

H

Hierarchical linear modeling, 127,
    130–133
    random effects, 133–134
    tau matrix, 133–134
    using R software, 134
        cross-level interactions, 136–138
        null model using achieve data,
            134–135
        random coefficients model, 136–138
        random intercepts models, 135–136
{Hmisc} package, 21–22
Homoscedasticity assumption, 42

I

Interaction, 61, 65
Intercept differences, 97

K

Kendall tau, 18
Kurtosis statistics, 48

L

Linearity assumption
    multiple regression analysis, 42
    Pearson r, 16
Linear regression equation, 28–30
lm.beta() function, 37
lm() function, 35

M

Mask attitudes, 2
Mask dataset, 79
Matrix scatterplot, 22
    using package, 22–23
Mediation, 102–103
    Baron and Kenny (1986) requirements
        for testing, 103
    example, 104–107

indirect effect, significance tests, 106
Model comparisons, 79–81
    of nested models, 84–85
        types, 85–86
    of non-nested models, 82
        R example, 82–84
Model fit information, 31
Moderation, 93–94
    R example, 94–95
Multicollinearity, 43
Multiple R, 31
Multiple regression analysis, 33, 41, 108
    absence of multicollinearity, 43–44
    categorical variables, 55–57
    checking assumptions, R software,
        44–45
        assessing presence of
            multicollinearity, 51–53
        homoscedasticity, 47–51
        independence, 47–51
        linearity, 45–46
        normality, 47–51
    dummy variables (*see* Dummy
        variables)
    example output, 34
    interaction effects, 61–65
    interpretational considerations, 42–43
    regression model, theoretically strong,
        43
    restriction of range, 43
    statistical assumptions, 41–42
Multivariate normality assumption, 42

N

Nested data, 127
Nested regression models, 81
Non-linearity, 111–112
    negative values, 113
    non-linear terms, 116
        example, 117–119
        multicollinearity, 117
        pros and cons, 119
    transformation selection, 112–113
        motivating example, 114–116
        pros and cons, 113–114
    variable transformations, 112
Non-parametric analysis

ranking across all groups, 153
of Spearman Rho and Kendal Tau, 152
ranking within variable, 152
Non-parametric correlations, 17
using cheating data, 19–20
Normality assumption
multiple regression analysis, 42
Pearson *r*, 15

O

Ordinary least squares (OLS) regression,
28, 35, 44
using lm(), 35–37
Outcome variable, 28
Overfitting, 116

P

Pearson product moment correlation
coefficient, *see* Pearson *r*
Pearson *r*, 10, 11
assumptions, 14
independence, 14–15
linearity, 16
normality, 15
scale of measurement, 15–16
correlations, 16–18
significance testing, 11
Perfectionism data, 36
{Performance Analytics} package, 22–23
plot(esteem.beta), 49
Predictive analysis, 3
Predictor variable, 28

R

R^2 change, 84
R^2 model, 31
Random assignment, 4
Rank ordering data, 151
R Commander, 142
Reference category, 58
Regression analysis, 3
as explanation, 3–4
as prediction, 3
Regression discontinuity, 96–97
interpreting treatment effects, 97

motivating example, 97–102
terminology, 97
Regression extensions 1
mediation
Baron and Kenny (1986)
requirements for testing, 103
example, 104–107
indirect effect, significance tests,
106
moderation, 93–94
R example, 94–95
regression discontinuity, 96–97
interpreting treatment effects, 97
motivating example, 97–102
terminology, 97
Regression extensions 2
cross-validation, 120–121
coefficient stability example,
123–125
procedures, 121–122
R cross-validation R squared,
122–123
samples, 121
non-linearity, 111–112
negative values, 113
non-linear terms, 116–119
transformation selection, 112–116
variable transformations, 112
Regression extensions 3
fixed effects modeling, 128
R using PT_FEM data, 128–130
hierarchical linear modeling, 130–133
HLM using R software, 134
random effects, 133–134
tau matrix, 133–134
Regression model fit, 30–31
Residuals, 47
Residuals plot, 49
Restriction of range, 16, 43
R function and package index
assumptions, 157
correlation and regression, 156–157
functions, 155–156
interactions, 157
regression extensions, 157–158
R shells
R software, 141–142
R software, 24, 141

centering predictors, 65–67
loading/importing data, 146–148
operations, 144–145
R console and scripts, 143–144
R help, 149
R packages, 142–143
R shells, 141–142
statistical functions, 145
R Studio, 141

Standard error of the estimate (SEE), 32
{stats} package, 20–21
Stepwise regression, 85, 86
    R example, 88–89
summary() function, 35
Sum of squared deviations, 28
Sum of squares between, 70
Sum of squares error, 70
Sums of squares total, 70

S

Scale of measurement assumption
    multiple regression analysis, 42
    Pearson r, 15–16
Scatterplot, 7, 16
    of positive and negative relationship, 8
SEE, *see* Standard error of the estimate
Sequential regression, 85
    R example, 86–87
Simple linear regression, 27
    using test anxiety data, 32–33
Skewness, 48
Slope coefficient, 29
Slope differences, 97
Sobel test, 106
Spearman rho, 17–18

T

Total effect, 103

V

Variance, 10
Variance inflation factor (VIF), 52
Visualizing relationships, 7–9

Y

Y-intercept, 29

Z

'0 0' category, 58

Printed in the United States
by Baker & Taylor Publisher Services

Printed in the United States
by Baker & Taylor Publisher Services